DEFINING A DECADE

Envisioning CSTB's Second 10 Years

Proceedings of CSTB's 10th Anniversary Symposium
May 16, 1996 — Washington, D.C.

Computer Science and Telecommunications Board
Commission on Physical Sciences, Mathematics, and Applications
National Research Council

NATIONAL ACADEMY PRESS
Washington, D.C. 1997

NATIONAL ACADEMY PRESS • 2101 Constitution Avenue, N.W. • Washington, DC 20418

NOTICE: This report derives from CSTB's core program, which was approved by the Governing Board of the National Research Council, whose members are drawn from the councils of the National Academy of Sciences, the National Academy of Engineering, and the Institute of Medicine.

This report has been reviewed by a group other than the authors according to procedures approved by a Report Review Committee consisting of members of the National Academy of Sciences, the National Academy of Engineering, and the Institute of Medicine.

The National Academy of Sciences is a private, nonprofit, self-perpetuating society of distinguished scholars engaged in scientific and engineering research, dedicated to the furtherance of science and technology and to their use for the general welfare. Upon the authority of the charter granted to it by the Congress in 1863, the Academy has a mandate that requires it to advise the federal government on scientific and technical matters. Dr. Bruce Alberts is president of the National Academy of Sciences.

The National Academy of Engineering was established in 1964, under the charter of the National Academy of Sciences, as a parallel organization of outstanding engineers. It is autonomous in its administration and in the selection of its members, sharing with the National Academy of Sciences the responsibility for advising the federal government. The National Academy of Engineering also sponsors engineering programs aimed at meeting national needs, encourages education and research, and recognizes the superior achievements of engineers. Dr. William A. Wulf is president of the National Academy of Engineering.

The Institute of Medicine was established in 1970 by the National Academy of Sciences to secure the services of eminent members of appropriate professions in the examination of policy matters pertaining to the health of the public. The Institute acts under the responsibility given to the National Academy of Sciences by its congressional charter to be an adviser to the federal government and, upon its own initiative, to identify issues of medical care, research, and education. Dr. Kenneth I. Shine is president of the Institute of Medicine.

The National Research Council was organized by the National Academy of Sciences in 1916 to associate the broad community of science and technology with the Academy's purposes of furthering knowledge and advising the federal government. Functioning in accordance with general policies determined by the Academy, the Council has become the principal operating agency of both the National Academy of Sciences and the National Academy of Engineering in providing services to the government, the public, and the scientific and engineering communities. The Council is administered jointly by both Academies and the Institute of Medicine. Dr. Bruce Alberts and Dr. William A. Wulf are chairman and vice chairman, respectively, of the National Research Council.

Support for this project was provided by core funds of the Computer Science and Telecommunications Board. Core support for CSTB is provided by its public and private sponsors: the Air Force Office of Scientific Research, Defense Advanced Research Projects Agency, Department of Energy, National Aeronautics and Space Administration, National Library of Medicine, National Science Foundation, Office of Naval Research, Apple Computer, Inc., AT&T Laboratories, Digital Equipment Corporation, Hewlett-Packard Company, Intel Corporation, Lucent Technologies, Inc., and Motorola, Inc. Any opinions, findings, conclusions, or recommendations expressed in this material are those of the symposium presenters and do not necessarily reflect the views of the sponsors.

International Standard Book Number 0-309-05933-X

Additional copies of this report are available from:

Computer Science and Telecommunications Board
2101 Constitution Avenue, NW
Washington, DC 20418
www2.nas.edu/cstbweb

Copyright 1997 by the National Academy of Sciences. All rights reserved.

Printed in the United States of America

CSTB'S 10TH ANNIVERSARY SYMPOSIUM PLANNING COMMITTEE

MICHAEL L. DERTOUZOS, Massachusetts Institute of Technology
LEONARD KLEINROCK, University of California, Los Angeles
JOSEPH F. TRAUB, Columbia University
WILLIAM A. WULF, University of Virginia

Staff

MARJORY S. BLUMENTHAL, Director
JEAN E. SMITH, Program Associate

COMPUTER SCIENCE AND TELECOMMUNICATIONS BOARD

DAVID D. CLARK, Massachusetts Institute of Technology, *Chair*
FRANCES E. ALLEN, IBM T.J. Watson Research Center
JAMES CHIDDIX, Time Warner Cable
JEFF DOZIER, University of California at Santa Barbara
A.G. FRASER, AT&T
SUSAN L. GRAHAM, University of California at Berkeley
JAMES GRAY, Microsoft Corporation
BARBARA J. GROSZ, Harvard University
PATRICK M. HANRAHAN, Stanford University
JUDITH HEMPEL, University of California at San Francisco
DEBORAH A. JOSEPH, University of Wisconsin
BUTLER W. LAMPSON, Microsoft Corporation
EDWARD D. LAZOWSKA, University of Washington
MICHAEL LESK, Bellcore
DAVID LIDDLE, Interval Research
BARBARA H. LISKOV, Massachusetts Institute of Technology
JOHN MAJOR, QUALCOMM Incorporated
DAVID G. MESSERSCHMITT, University of California at Berkeley
DONALD NORMAN, Hewlett-Packard Company
RAYMOND OZZIE, Iris Associates, Incorporated
DONALD SIMBORG, KnowMed Systems
LESLIE L. VADASZ, Intel Corporation

MARJORY S. BLUMENTHAL, Director
HERBERT S. LIN, Senior Staff Officer
JERRY R. SHEEHAN, Program Officer
ALAN S. INOUYE, Program Officer
JON EISENBERG, Program Officer
MARK E. BALKOVICH, Research Associate
LESLIE M. WADE, Research Assistant
LISA L. SHUM, Project Assistant
SYNOD P. BOYD, Project Assistant

COMMISSION ON PHYSICAL SCIENCES, MATHEMATICS, AND APPLICATIONS

ROBERT J. HERMANN, United Technologies Corp., *Co-chair*
W. CARL LINEBERGER, University of Colorado, *Co-chair*
PETER M. BANKS, Environmental Research Institute of Michigan
LAWRENCE D. BROWN, University of Pennsylvania
RONALD G. DOUGLAS, Texas A&M University
JOHN E. ESTES, University of California at Santa Barbara
L. LOUIS HEGEDUS, Elf Atochem North America Inc.
JOHN E. HOPCROFT, Cornell University
RHONDA J. HUGHES, Bryn Mawr College
SHIRLEY A. JACKSON, U.S. Nuclear Regulatory Commission
KENNETH H. KELLER, University of Minnesota
KENNETH I. KELLERMANN, National Radio Astronomy Observatory
MARGARET G. KIVELSON, University of California at Los Angeles
DANIEL KLEPPNER, Massachusetts Institute of Technology
JOHN KREICK, Sanders, a Lockheed Martin Company
MARSHA I. LESTER, University of Pennsylvania
THOMAS A. PRINCE, California Institute of Technology
NICHOLAS P. SAMIOS, Brookhaven National Laboratory
L.E. SCRIVEN, University of Minnesota
SHMUEL WINOGRAD, IBM T.J. Watson Research Center
CHARLES A. ZRAKET, MITRE Corp. (retired)

NORMAN METZGER, Executive Director

Preface

As the Computer Science and Telecommunications Board (CSTB) approached its 10th anniversary in 1996, a number of current and former members began to entertain the idea of convening a special event to celebrate the occasion. Taking the lead were the first chair of CSTB, Joseph Traub; then chair, William Wulf; and former Board members Michael Dertouzos and Leonard Kleinrock. The planning committee expanded the customary afternoon open session of the Board's spring meeting into a one-day symposium on May 16 in Washington, D.C. Two goals were identified for the event: to celebrate 10 years of CSTB's achievements and to begin to define its goals for the next decade.

In order to involve as many CSTB alumni as possible, the committee enlisted many former Board and committee members as presenters. In addition, two panels were convened: one to address ways in which CSTB interacts with its sponsors and one to identify the emerging issues likely to be on CSTB's agenda during the next 10 years. To expand the audience for these exciting presentations and to broaden the discussions begun on May 16, 1996, the proceedings have been compiled. This volume represents the edited versions of the presentations, which were also reviewed by several CSTB alumni for accuracy and meaning. It reflects the events of a day that was characterized by thoughtful presentations and lively discussion. Although some speaker affiliations have changed since the symposium, in general, those in effect then are presented to preserve the context for the discussion. The groundwork was laid for setting CSTB's agenda for the next 10 years, opening the door for the Board's further thought and refinement over the following months.

This publication is part of a body of publications produced by the Computer Science and Telecommunications Board in its 11-year history. The Board, part of the National Research Council within the National Academy of Sciences, was established in 1986 to provide independent guidance to the federal government on technical and public policy issues in computing and communications. Composed of leaders from industry and academia, CSTB conducts studies that recommend actions by government, industry, and academic researchers on critical national issues. These studies, prepared by principals from the public and private sectors, provide a balanced perspective on the issues at hand. CSTB also provides a neutral meeting ground for the consideration and focusing of complex issues where resolution and action may be premature. It convenes invitational discussion sessions that bring together principals from the public and private sectors to share perspectives on all sides of an issue, ensuring that the debate is not dominated by the loudest voices.

CSTB is grateful to the symposium presenters, moderators, and participants for their contributions. As always, special thanks go to CSTB's sponsors, whose ongoing support made this symposium possible. These include the Air Force Office of Scientific Research, Defense Advanced Research Projects Agency, Department of Energy, National Aeronautics and Space Administration, National Library of Medicine, National Science Foundation, Office of Naval Research, Apple Computer Inc., AT&T Laboratories, Digital Equipment Corporation, Hewlett-Packard Company, Intel Corporation, Lucent Technologies Inc., and Motorola Inc. CSTB also wishes to thank SEMATECH for a contribution that helped to defray symposium costs.

DAVID D. CLARK, *Chair*
Computer Science and Telecommunications Board

Contents

1. FROM INFOWARE TO INFOWAR 1
   Joseph F. Traub

2. LINKING THE CSTB COMMUNITY TO THE FEDERAL GOVERNMENT:
   EXPERT ADVICE FOR POLICYMAKERS 8
   Michael R. Nelson (Moderator), David B. Nelson, Paul R. Young, and Howard Frank

3. THE GLOBAL DIFFUSION OF COMPUTING: ISSUES IN DEVELOPMENT AND POLICY 18
   Seymour E. Goodman

4. ENGINES OF PROGRESS: SEMICONDUCTOR TECHNOLOGY TRENDS AND ISSUES 22
   William J. Spencer and Charles L. Seitz

5. COMPUTING AND COMMUNICATIONS UNCHAINED: THE VIRTUAL WORLD 36
   Leonard Kleinrock and John Major

6. PICTURE THIS: THE CHANGING WORLD OF GRAPHICS 47
   Henry Fuchs, Donald P. Greenberg, and Andries van Dam

7. COMPUTATIONAL BIOLOGY AND THE CROSS-DISCIPLINARY CHALLENGES 53
   Federal Research and Funding Policies, *Deborah A. Joseph*
   Finding a Home in Academia, *Edward H. Shortliffe*

8. VISIONS FOR THE FUTURE OF THE FIELDS 63
   *David D. Clark (Moderator), Edward A. Feigenbaum, Juris Hartmanis,
   Robert W. Lucky, Robert M. Metcalfe, Raj Reddy, and Mary Shaw*

9. UNIQUE CHALLENGES: COMPUTING AND TELECOMMUNICATIONS
   IN A KNOWLEDGE ECONOMY 73
   Ellen M. Knapp

10 ANCIENT HUMANS IN THE INFORMATION AGE 83
    Michael L. Dertouzos

APPENDIXES

A   Letter from Dr. Bruce Alberts, President, National Academy of Sciences 89
B   Symposium Attendees 91
C   Biographies of Presenters 96

1

From Infoware to Infowar

Joseph F. Traub

At its first meeting on May 7 and 8, 1986, the Computer Science and Technology Board (CSTB) identified six critical national issues. I will tell you how those issues looked in 1986 and how they look today from my 1996 perspective. Then I will turn to a seventh issue that I did not want to raise publicly in 1986.

From its inception, CSTB deemed it important to look at both technical and policy issues. In 1986, it was fairly unusual for National Research Council (NRC) boards to consider policy issues. Therefore, at that first meeting we invited senior federal officials to give their views of the critical national issues. The officials included Congressman Donald Fuqua, chairman, U.S. House of Representatives Committee on Science and Technology; John McTague, acting science adviser to President Reagan; Gordon Bell, assistant director, computer and information science and engineering, National Science Foundation; Robert C. Duncan, director, Defense Advanced Research Projects Agency (DARPA); and Alvin Trivelpiece, director of research, U.S. Department of Energy.

I will discuss the six critical areas identified in May 1986. When I speak about them from my 1996 vantage point, I am giving only my own view; I have not had the benefit of discussing these opinions with the Board.

In May 1986, the Board's number one concern was competitiveness. *How can the United States best ensure the continued leadership of its computer science and technology enterprise in the face of intensified global competition?*

Competitiveness was a big issue in 1986. Indeed, the opening sentences of the NRC's press release announcing formation of the Board were these: "U.S. leadership in computer research and manufacturing has been seriously eroded. The NRC has established a computer science and technology board to advise federal agencies and private firms on ways to strengthen U.S. international competitiveness in this field and to ensure that the full promise of this area is realized." We were deeply concerned about our competitive position, particularly vis à vis the Japanese, in areas such as computer chips, artificial intelligence, and software. For example, we had heard that software from Japanese factories had unusually few errors.

Today, we must remain vigilant regarding our competitive position. The economic and national security stakes are higher than ever, but our worst fears of 1986 were not realized. For example, according to William Spencer, president of SEMATECH, in the mid-1980s we were losing 3 to 4 percent of market share per year. Today, the United States and Japan have about 40 percent of market share each, while the rest of the world has 20 percent of the $150 billion semiconductor market. Business is booming for American companies in software, semiconductors, Internet-related infoware, and above all, in producing content. We must maintain and leverage these positions in an economy that is increasingly international.

The second item on CSTB's list was talent. *How can the gap be closed between the small number of U.S. citizens graduating with Ph.D. degrees in computer science and computer engineering and the large demand for graduates with these skills? How can high school graduates be taught to deal with the computers and high technology they must use on the job in fields that range from the military to banking?*

I will divide this issue into two parts: Ph.D. production and K-12 education. We have succeeded in building Ph.D. production from some 200 a year to about 1,000 a year in computer science and engineering. In the 1990s, there has been concern about overproduction of Ph.Ds in some specialties as universities and research laboratories downsize and as the field matures. I am not convinced the country needs 1,000 Ph.D.s a year in computer science and engineering. I have seen the pain in physics and mathematics since about 1970; can *we* learn from history? Incidentally, after a quarter of a century, and due to a number of circumstances, the crunch in physics and mathematics seems worse than ever. Since most of these Ph.Ds will not be appointed to faculty positions in the top research universities, questions have also been raised about possible changes in the Ph.D. program to produce people better suited to positions in industry and colleges.

Problems in K-12 education seem more serious and overwhelming than ever. There is a widespread feeling, which I certainly share, that Americans do not get the education required for an informed populace in a democracy. They do not get an education that will enable them to fill many service sector jobs or to function in our high-technology armed forces. Students are not adequately prepared in analytic and writing skills to do well in universities. If we believe that educated people will be the key resource in the twenty-first century, we have much to be concerned about. Bruce Alberts, president of the National Academy of Sciences (NAS), has identified K-12 education as the most important problem of his presidency.

The third item on the 1986 list was scope and support. *What will be the nature of computer science and technology in the 1990s? How can its health and vitality be sustained during a period of uncertainty and stringency in federal research and development budgets?*

I will discuss this issue in two parts also, beginning with nature and scope. It is typical of NRC boards to do studies on the nature of their fields. Some have become classics, such as the Bromley report for physics[1] and the Pimentel report for chemistry.[2] These studies set the standard for other studies. Since we were the "new kids on the block," CSTB decided that beginning with a report on the nature of the field would be self-serving. We wanted first to build a record of reports dealing with critical national issues. *Computing the Future*, CSTB's study of the scope and direction of computer science and technology, was published in 1992. The committee was chaired by Juris Hartmanis.

The second part of this item asked, *How can the field's health and vitality be sustained during a period of uncertainty and stringency in federal research and development budgets?*

Oh, my prophetic soul! Since there are people in this audience who have been grappling with this issue during a period of uncertainty and stringency, I will not pursue it here.

Item 4 on the 1986 list was supercomputers. *How can the power of supercomputers be exploited to promote scientific and technological advances, and how can U.S. leadership in this area be maintained?*

Supercomputers continue to have economic and symbolic importance. At the time that I write this paper, a major battle is under way to determine whether the National Center for Atmospheric Research will purchase an American or a Japanese supercomputer. The contract is valued at $13 million to $35 million; the amount of interest being generated must be due to more than just the amount of money involved. A teraflop computer will soon be installed at Sandia National Laboratories. Faster, more powerful machines are needed for aircraft design to ensure the safety and effectiveness of nuclear weapons with zero testing, for molecular dynamics in biology, and for cosmological computations.

The federal government's interest in supercomputers has evolved since 1986. A program plan for high-performance computing was published by the Office of Science and Technology Policy (OSTP) in 1989. The High Performance Computing Act of 1991 authorized a five-year program known as the High Performance

---

[1] Board on Physics and Astronomy, National Research Council. 1972. *Physics in Perspective*. National Academy Press, Washington, D.C.
[2] Board on Chemical Sciences and Technology, National Research Council. 1985. *Opportunities in Chemistry*. National Academy Press, Washington, D.C.

Computing and Communications Initiative (HPCCI). In 1995, a CSTB study, chaired by Frederick Brooks and Ivan Sutherland, responded to a congressional request for an assessment of HPCCI with a study titled *Evolving the High-Performance Computing and Communications Initiative to Support the Nation's Information Infrastructure.* Although high performance continues to be extremely important, there is less focus on supercomputers since the parallel processing paradigm has been accepted and higher performance has become pervasive.

Issue 5 was software. *What can be done to promote the economical production of reliable software, which represents a major portion of the cost and effort in the design and use of new computer systems?*

I was looking through my CSTB files and found the following note from 1987: "Sam Fuller says Bill Gates has really thought about software; he would be a good person to talk to the Board." Quite so, Sam.

In April 1988, CSTB met at Stanford University rather than in Washington, D.C. Edward Feigenbaum arranged for John Young, president and chief executive officer of Hewlett-Packard, to host a breakfast at which leaders of Silicon Valley companies could tell us what national issues most affected them. Attending the meeting was the chairman of Hambrecht and Quist, one William Perry. I was struck by the consensus among these industrial executives on the two most important issues. They identified K-12 education because they were concerned about finding good employees, and they identified software because they recognized that the economical production of reliable software was a crucial problem for their businesses. This meeting underscored the Board's concerns about software.

The final item on the initial list was infrastructure. *What are the important underlying capabilities, such as national networks and electronic libraries, that are needed to support the healthy evolution of computing? How can they be provided in a timely and effective fashion and integrated into daily activities to enhance national productivity?*

The first two major studies the Board published were *Toward a National Research Network* in July 1988 and *The National Challenge in Computer Science and Technology* in September 1988. *The National Challenge* was unique in that the study was done by the Board rather than by a committee specifically appointed for the purpose. Since it was a Board report, we were all equally involved. However, Michael Dertouzos was far more equally involved than the rest of us! *The National Challenge* presented two recommendations. Here is the first in its entirety:

> Enhanced, nationwide computer networking should be seen as essential to maximizing the benefits in productivity and competitiveness that are created by computers. Networking will facilitate the application and delivery of diverse advances in computer science and technology to the benefit of all segments of society. The board envisions an enhanced national information networking capability, and it has already begun to examine a host of related questions about how physically to improve data networking infrastructure; associated costs, impacts, and benefits; and the roles of industry, government, and other interested parties.

Seems rather prescient from the vantage point of 1996.

During the past 10 years, infrastructure certainly became prominent in the Board's portfolio. Indeed, CSTB was initially the abbreviation for Computer Science and Technology Board. After the Board was asked by Frank Press, then the president of the NAS and chairman of the NRC, to add telecommunications to its responsibilities, Marjory Blumenthal and I decided to rename it the Computer Science and Telecommunications Board to preserve the abbreviation.

The Board's list of critical issues in 1996 can be ascertained by looking at the list of current and future initiatives listed in CSTB's new brochure. You might want to compose your own list of the six most critical national issues in computing and telecommunications in 1996.

In 1986, there was also a seventh topic on my mind. In 1985, I began to notice various ways in which the information infrastructure was vulnerable to electronic or physical attack. I imagined myself to be a terrorist or an enemy country, and targeted aspects of what today we would call "the national information infrastructure" (NII). I did not go public with this concern because I feared it would do more harm than good. There is now considerable attention being paid to our vulnerability, especially by the Department of Defense and the media. However, I have strong concerns, and I feel now is the time to express them. I will focus on the civilian infrastructure, although it is sometimes difficult to separate military from civilian in this domain.

It is just because the United States is the most advanced country in the world when it comes to the use of information technologies that we are also the most vulnerable to attack. I will give you one illustration. Financial security is obviously very important to us as a people. We have two primary types of assets: real estate and what I will call *virtual estate.* I will confine my comments to virtual estate, which consists of bank accounts, equities, certificates of deposit, pension accounts, and so on. As you all know, this is "virtual" estate because it is recorded in electrons. If you were a terrorist and wanted to do a great deal of damage to American institutions and individuals, a natural target would be the virtual estate. Although there are electronic backups and paper trails, I am not convinced that virtual estate is secure.

Then there are the electrons in the foreign exchange markets. According to the Bank for International Settlement, turnover in this market is $1.2 trillion a day. This, of course, overwhelmingly dominates the value of goods moved around the world. Furthermore, an amount of money that equals the annual gross national product of the United States moves through the foreign exchange markets in not much more than a week—and it all consists of electrons. These markets are international, but a successful attack could also have major domestic consequences; the line between an international and a domestic attack has become blurred. This fading of distinctions is characteristic; we will see more examples later.

Our virtual estate is just one example of a potential target. Others include the power grid, the air traffic control system, and the communications system. I see the protection of information assets as a national security issue, although this view is not universally shared. In November 1993, I was one of seven civilians who participated in a seminar convened by Andrew Marshall, director, Net Assessment, in the Office of the Secretary of Defense. Marshall is a highly respected and influential pioneer in what is sometimes called Revolution in Military Affairs (RMA). Mr. Marshall organized the meeting because he believed that the world stands on the threshold of new information technologies that would have a profound effect on the U.S. military establishment. Participating on the military side was a group of general officers from all the services.

I argued that there was a history of the military defending transportation assets—such as rail lines, harbors, rivers, and airports—during wartime. Should the armed services consider protecting the national information infrastructure? Such a mission poses many questions and presents its own risks. It is further complicated by the blurring of wartime and peacetime activities that has already begun and will likely increase.

There was a surprising split in views among the participants. A number of civilians, myself included, felt that the protection of civilian information assets might fall within the parameters of national security, whereas the military believed that responsibility for protecting these assets belongs in the realm of private industry and the police. Military uses of infowar (IW), both defensive and offensive, are clearly the responsibility of the Department of Defense (DOD). In January 1995, the Secretary of Defense established the IW Executive Board to facilitate "the development and achievement of national information warfare goals."[3]

I want to focus here on protection of civilian assets, which, unfortunately, can be difficult to separate from military assets. An enemy might attack the United States exactly by attacking the civilian infrastructure. This is just an example of a more general tendency. In his book *The Transformation of War*, Martin van Creveld argues that the low-intensity conflicts that have become the norm since World War II will become far more prevalent and will spread to developed countries, instead of being confined primarily to the Third World. He writes: "As the spread of low-intensity conflict causes trinitarian structures to come tumbling down, strategy will focus on obliterating the existing line between those who fight and those who watch, pay, and suffer." By trinitarian structures, van Creveld means the division among the state, the military, and the people.[4]

Here are two specific examples of the difficulty in separating military assets from civilian assets. Approximately 95 percent of all military communications are routed through commercial lines. We buy most of the chips used in military systems from commercial vendors, many of whom are located in foreign countries. Why might

---

[3]To learn more, see Berkowitz, B. 1995. "Warfare in the Information Age," *Issues in Science and Technology*, Fall, pp. 59-66; and Molander, R.C., A.S. Riddile, and P.A. Wilson. 1996. *Strategic Information Warfare.* National Defense Research Institute, RAND, Santa Monica, Calif.

[4]van Crevald, M. 1991. *The Transformation of War.* The Free Press, New York.

infowar be the weapon of choice of foreign or domestic terrorists, and of foreign countries, small and large? Here is a partial list of reasons:

- The United States is the only current superpower. As the Gulf War showed, it is foolish to challenge us in conventional war.
- Since we have the most advanced NII, it is the most vulnerable to attack. (Of course, this would give us an advantage in offensive IW.)
- The price of entering into IW is low.
- The learning curve is very steep.
- It might be difficult to detect who attacked us or even whether there was malicious intent.

Who should be in charge of protecting the civilian infrastructure against attack? I believe that there needs to be strong government leadership and that it should be located in the executive branch. Should the lead role be played by an existing entity, a combination of existing entities properly coordinated, or a government structure created for this purpose? This is an exceptionally complex and important question that I will not pursue here.

It has been argued that the NII will acquire at least partial immunity due to repeated attacks, analogous to a biological organism. It has also been argued that the problem can be left to the private sector to solve. Although I believe that the private sector has a very important role to play, I am not convinced it can do this on its own. Coordination between the government and the private sector will be another difficult and important area to address.

Protecting ourselves against infowar may require the careful balancing of our desire for liberty and privacy with our wish for security. As the Clipper Chip illustrates, the conflicting demands of privacy, commerce, and security can generate strong tensions. It is a particularly difficult issue for democracies, far more difficult than for countries that do not place a high value on privacy and liberty.

How imminent is a serious IW attack? How much time do we have to prepare? Unlike most conventional warfare, in which there is a visible buildup of forces, an IW attack may come without warning. Although we should think carefully about how to meet the threat, I believe that time is of the essence. I suggest that no issue is more important to the nation than the defense of our national information infrastructure.

DISCUSSION

JOHN MAJOR: Do you see a dichotomy between the military's desire, on the one hand, to restrict the level of security that private industry has access to, and on the other, to pass on to private industry the responsibility for protecting its information from international attack?

JOSEPH TRAUB: It is a minefield of difficulties. Some of the major companies, such as AT&T and MCI, for example, say, "Do not worry, we can take care of it." You know how difficult it has been to conduct the Board's study on encryption. I have no special wisdom on this subject, but I am sure there are people in the room who do.

HOWARD FRANK: I would like to react to the original question because it contains an assumption that is not necessarily true—that the military has a desire to reduce or restrict the amount of protection afforded to the civilian economy and infrastructure. I do not think this is true. There has been an ongoing debate, which is certainly not uniform throughout DOD, let alone among the major political leaders, about certain export policies and so on. In general, there is no firm policy that says that the civilian sector should have less security or protection than DOD.

In the first place, I think the problem has been one of technology not responding to the needs because they are basically invisible. We are more aware of this now. Maybe you should have said something in 1986 or 1987, because it appears to me that this came out of the blue. For instance, if you look at the research community, it is devoid of original ideas on this aspect of the problem. It is only now that people are beginning to think about what we might do as a nation in terms of a research agenda.

Second, the issue of privacy is one of the great perplexities in law enforcement. Resolving this issue is going to take quite a while, but it seems to be moving in the right direction across the government.

Third, the civilian sector, industry by industry, varies greatly in terms of what it believes the vulnerabilities are. In fact, I participated in an NRC study in 1989-1990 that looked at the security preparedness of the telephone system. When we began that study, there was vast disagreement about whether there were any vulnerabilities whatsoever. Then, in the middle of the study, there was a major service disruption in Hinsdale, Illinois, and a failure in Bedford, Massachusetts, followed by another failure. So the reaction was, "Oh, well, maybe we really made a mistake. Maybe we are really vulnerable."

Other parts of the infrastructure, like the power industry, have their heads buried in the sand; there is virtually no security at all. So this is a very complex question, and it is not a dichotomy. It is a continuum of issues, many of which are economic. For instance, given the fact that we do not have a nationwide server for infrastructure, let us say telecommunications, no major trade can now be made that says this is in the national interest and therefore you must do it—the way we could do in providing universal telephone service 50 years ago. This complicates the problem a lot.

SIDNEY KARIN: I want to thank you for raising the information warfare issue. I think it is absolutely critical. You said that determining who should be in charge is very complex. I think that answering the question of whether anybody should be in charge is down the road. We have not gotten there yet.

I think the threat is real. I think in some nontrivial way we are under attack today. There is lots of evidence that security breaches have been taking place for the past several years. The major problem is that we, as a society, have not recognized that there is a threat, a real danger, and that the consequences could be quite serious.

So I submit that the issue is not just a quibbling over civilian versus military security. Although I have a very strong position on this issue, I will not get into it at the moment. The issue is that, as a society, we have not agreed that security is a serious problem. Until we agree on this, there will not be any solutions—no matter how hard anybody works at trying to implement anything, no matter what structures are imposed by anybody to deal with it.

The first thing that needs to be done is to raise everybody's consciousness that something bad could happen, and that there are people trying to make it happen for various reasons. I commend you for raising the issue.

BUTLER LAMPSON: The basic fact about security is that it is expensive; it is a pain in the neck; and people are going to implement it only in response to convincing evidence that there is a problem. The only way you are going to get that evidence is when someone you think is in a position similar to yours gets badly hurt.

In my view, all this discussion about what we ought to do and how much we ought to worry about the threat is entirely beside the point because nobody is actually going to do anything until there is some serious damage done. The fact is that no serious damage has been done. It is fine for us to think about how we might respond once the motivation is there to respond, but I think trying to raise the level of motivation is completely pointless.

KARIN: What would constitute evidence in your judgment that somebody has been badly hurt? What would it take for your organization to decide that one of its competitors was badly hurt? What is a hypothetical incident that would convince you?

LAMPSON: Something that happens that costs you a lot of money. Bankruptcy would do it. That would definitely have an impact.

KARIN: How large a company would have to go bankrupt before your company would take notice, recognize there is a serious threat, and decide to change its behavior?

LAMPSON: How large a company would have to go bankrupt? I do not know. It is an interesting question. Since no companies have gone bankrupt yet, right now the question is academic. If you believe that some companies have gone bankrupt as a result of information warfare and you want to promote action, then I strongly urge you to find out who those companies are and to publicize the situation clearly. This will have far more effect than anything else that you or Joe or anyone else might say.

ROBERT BONOMETTI: I would like to make the suggestion that the problem actually is more extensive than just a deliberate, determined attack against the infrastructure. As society becomes more dependent on this infrastructure, natural disasters also become of great concern because the social fabric will be ripped apart and become dysfunctional in the aftermath of a disaster.

There are examples we can look at, such as the aftermath of Hurricane Andrew in Florida, that provide some lessons learned. One very trivial but interesting fact is that people were unable to get cash because everyone was so dependent on ATMs (automatic teller machines). After the hurricane the infrastructure was not there, and it was a problem for some time.

ANDRE VAN TILBORG: I wonder if you might comment on how much of the security problem might be related to not having the analogue of something like building codes for information and telecommunications systems, versus how much really requires deep, new insights and research. If you had, for instance, a national electrical code—not for electricity or electrical appliances or Underwriters Laboratory, but for computing systems and telecommunications systems—you might be able to cover a large fraction of the troublesome areas in ensuring that your systems remain stable and work. You would still have that part where a very determined adversary can get through, even though you have a good building code. I wonder what your thoughts on that might be.

TRAUB: I think what you are suggesting is standards. It has been suggested that there be a core communications system in case of a national emergency, and there has been some discussion about that. The things we have with which to protect our homes will at least keep out the amateurs, although they will not keep out a professional, determined burglar. The big problem is that everything is changing so quickly that this is an almost impossible area, I believe, to standardize. Standardization, it seems to me, requires a certain maturity and a certain stability. I am not sure we can do that in this area.

ROBERT KAHN: The good news is that the notion of the national information infrastructure is in the public consciousness. The bad news is that we really do not know what it is or might not recognize it if we saw it. To some, the NII is 500 channels of cable TV and to others it is the Library of Congress on every desktop—in some ways very mutually incompatible goals.

Some people think we have always had an information infrastructure, or at least maybe since 1844 when the telegraph was invented, if you want to focus on electrons. Others think we clearly have it now. Still others will wake up in 10 years and be totally shocked that we do not have an infrastructure yet and have been talking about it all these years.

It seems to me that the one thing that has really been missing, apart from understanding what it is, is any notion about getting coherence among all the pieces, so that the infrastructure really becomes the mechanism to lower the barriers to productivity in a broad sense. I do not think we are there yet. The big objective over the next decade, perhaps many decades, is trying to figure out how to achieve interoperability to lower these barriers. By getting coherence in the system, you make it more of a target for the kind of information warfare that you are talking about.

So my question to you is, How do we go about designing this coherence for interoperability into the system, while at the same time worrying about protecting against the kind of information warfare you are discussing as a social process in this country?

TRAUB: That is a very good question, Bob. I am sorry, but my time is up.

WILLIAM WULF: Let me just make a couple of comments. First of all, the balancing of privacy and societal protection was mentioned. Probably the most sensitive report that CSTB has ever undertaken is going to be released imminently, and it is one that was requested by Congress.[5] It addresses national cryptography policy. It will be a completely unclassified report. This fact is very, very important. There is not going to be a classified annex to the report. We wanted the report to be completely unclassified.

My second comment has to do with the national information infrastructure and natural disasters. We absolutely agree with Dr. Bonametti's remarks. If you look in the CSTB brochure, you will see that one of CSTB's 1996 studies (*Computing and Communications in the Extreme: Research for Crisis Management and Other Applications*) is looking at how we can use information technology to save lives and property.

In his early remarks, Joe pointed out that, from the outset, the Board has been concerned with both technology and policy issues. If anything—and this is a personal perception—the increasing recognition among people in both the executive and the legislative branches of the relevance of information technology to virtually every problem that the country faces has reinforced the correctness of that original decision.

---

[5]Computer Science and Telecommunications Board, National Research Council. 1996. *Cryptography's Role in Securing the Information Infrastructure.* National Academy Press, Washington, D.C.

2

Linking the CSTB Community to the Federal Government: Expert Advice for Policymakers

Michael R. Nelson, Moderator
David B. Nelson
Paul R. Young
Howard Frank

Michael R. Nelson

I am Mike Nelson from the White House Office of Science and Technology Policy. I am one of the two people at the White House who work full-time on information technology policy issues. I think it is probably the second most enjoyable job at the White House, after the President's. He has the advantage of not having to report to anybody, except to the American people.

I am actually a geophysicist by training, and it is interesting that I have ended up where I am. I came to Washington from the Massachusetts Institute of Technology about eight years ago on a one-year fellowship. Apparently I contracted "Potomac Fever." I hope it is not terminal. It might seem odd that a geophysicist would be working in this area, but as I have told many of you, the training is actually perfect. Having a sense of geologic time in Washington is very important.

My career in Washington has nearly paralleled the existence of the Computer Science and Telecommunications Board (CSTB) in terms of time. We have had a very interesting, fruitful, and I would say symbiotic, relationship. In Washington, you must have a good mentor if you are going to accomplish anything. I have been blessed with several, and several of them are on the Board. In working on information technology issues during the past eight years, I have worked very closely with the Board. I probably have relied on CSTB's reports as much as anyone and have been able to accomplish a lot more because of the information provided to me by individual Board members and the reports that CSTB has produced.

When I first came to Washington, I started working on earthquake issues. I spent two or three months doing this and realized that these were really hard issues, and they were very depressing. I would start my morning by reading scenarios of a major earthquake in Los Angeles killing 20,000 people. So I started working on global warming issues, where I read scenarios in which 2 million people would die. Then I decided it was much more fun to work on information technology issues: we all know that information technology will solve all of our problems; so this is where I now spend all my time.

The first hearing I organized for Senator Gore, then chairman of the Science, Technology, and Space Subcommittee in the Senate, was on computer technology and high-speed networks. Robert Kahn, among others, testified. Leonard Kleinrock was our lead witness, testifying on CSTB's second report, *Toward a National Research Network* (1988). It was a very influential, very important report. I know it was influential because I took large portions and inserted them directly into the briefing memo, which was given to all the senators who came to the hearings. I also inserted parts of that report directly into the legislation that I drafted for Senator Gore. Luckily, plagiarism is legal on the Hill! After that first hearing, several people were astonished at Senator Gore's grasp of both the technology and the issues. This was due in part to the fact that he had read the report, which was released on the same day.

Since that time, I have worked with the Board on a number of critical issues. CSTB has helped us design the high-performance computing legislation and keep the High Performance Computing and Communications Initiative on track. It has helped us deal with issues such as computer security. The Board has helped us evolve the ARPANET into the NSFnet, into the National Research and Education Network, into the Internet, into the Net, into the national information infrastructure, into the global information infrastructure, and into whatever it is we are now creating.

Today, we are going to look at how the Board has influenced policy and how it has worked effectively on the interface between science and engineering and policy making. Being one of the denizens of the interface, I know it is a pretty turbulent, unpredictable place to work. I know that this Board has been very effective in informing and enlightening those of us who have tried to help the policy-making process along. Today we have an excellent panel, and I should say that I am impressed with the quality of the entire symposium program.

This panel is going to take a Dickensian approach to its presentation by looking at CSTB's past, present, and future. The "spirit of CSTB past" will be provided by David Nelson (no relation), who has been working in this area even longer than I have and has been one of CSTB's primary customers. The "spirit of CSTB present" will be provided by Paul Young, who will talk about what is going on now and some of the ways in which the Board is influencing policy making. The "spirit of CSTB future" will be represented by Howard Frank, since the Defense Advanced Research Projects Agency (DARPA) is always 10 years ahead of the rest of us. He will discuss what might be ahead for the Board. We are going to keep these remarks very short, about five minutes each, so that we can have a full discussion. All of the panelists can talk about the past, present, and future, and they will do so in the question-and-answer period. Our goal is to provide a chronological look at where the Board has been, where it is now, and what directions it might take in the future.

David B. Nelson

I will cover CSTB past. This does feel a little bit like a role in Charles Dickens's *A Christmas Carol*. As I was preparing this, I thought, "I am putting myself in the role of a peer reviewer." This reminded me of a story that I think most of you know, but it is so good that I will repeat it. This is the classic peer reviewer's report. It reads as follows: "This paper is novel, interesting, and correct. Unfortunately, the part that is novel is not interesting. The part that is interesting is not correct. The part that is correct is not novel." Fortunately, I do not make that judgment when it comes to the work of CSTB.

Before I review some of the Board's efforts from a government standpoint, let me remind you what the federal scene was like in 1986-1987. The Lax report[1] had been out for only a few years, and the various agencies were struggling to implement its recommendations. I viewed that as an attempt to cure the VAX disease. What I mean is that every chemistry department could afford a VAX computer; so the standard of excellence in chemistry computation was one VAX unit.

Next, DARPA had just begun its strategic computing initiative. Robert Kahn smiles from the audience—he remembers that one well. The National Science Foundation (NSF) was dealing with the question of establishing supercomputer centers, particularly how to link them together in something that would be called NSFnet. To remind you how history is out of our control, remember the study that said that the NSFnet should run Open Systems Interconnection (OSI) protocols as soon as practical. Think back. There is a lot of water under the bridge on that one.

---

[1]*Report of the Panel on Large Scale Computing in Science and Engineering,* Peter Lax, chairman, sponsored by the U.S. Department of Defense and the National Science Foundation, in cooperation with the U.S. Department of Energy and the National Aeronautics and Space Administration, Washington, D.C., December 26, 1992.

The Department of Energy (DOE) had responded to the Lax report by consolidating and extending all of its energy research computing, using what is now called the National Energy Research Supercomputer Center, and had just begun the ESnet that went with it. We were struggling with the question of what protocol to run. At just about that time, Cisco came along with multiprotocol routers that allowed us to sidestep this terrible political question—another instance in which technology saves you from your political fate.

The National Aeronautics and Space Administration (NASA) was planning the National Aeronautical Simulator, the numerical wind tunnel. A lot of universities were trying to figure out how their budgets could afford advanced computing. As we know, some could and some could not. On the Hill, as Michael Nelson has noted, Senator Albert Gore was thinking rather deeply about networking and about the things that networking could do for the country's future.

In the interagency setting, the Department of Defense (DOD), DOE, NSF, and NASA were working on the beginnings of what was then called the High Performance Computing Initiative. In the 1987 White House Office of Science and Technology Policy (OSTP) report, which is less well known than the 1989 report, it was referred to as high-performance computing. I believe it was Allan Bromley who said—as the 1989 report was being prepared to launch the HPC Initiative—"Shouldn't there be a second C? Isn't communications becoming pretty important?" As a result of such suggestions, it became known as the High Performance Computing and Communications Initiative (HPCCI).

With this introduction, let me review some of CSTB's work. In terms of quantity, CSTB has been very productive—more than 40 projects since 1987; approximately three to five substantial products per year, including one major study each year. Some of the substantial products included influential colloquia as well as long-term studies.

Next, CSTB has usually been ahead of the issues. I think Joseph Traub already mentioned this fact, and I would certainly agree. If we run very quickly through some of the products, we will see that it is true. As early as 1988, in *Toward a National Research Network*, CSTB addressed issues in the Gore bill and discussed the value of the research network.

Among CSTB's reports, there are several that resonated with me and that spoke particularly to policy issues. Many of you probably have a different list, so this is not intended to be complete or in any way a judgment about the ones that are not on my list.

In *Computers at Risk: Safe Computing in the Information Age* (1991), CSTB addressed the question of hackers and penetration. It did so in a somewhat low-keyed way because it considered system design and accidents to be as important as hackers. I would say that these are still issues. Many more people are killed in accidents than in murders and wars. So, while we worry about malicious attacks, our own systems are fairly fragile. *Computers at Risk* pointed out that inadvertent problems can bring things down as much as malicious penetration.

Keeping the U.S. Computer Industry Competitive was a series of reports (issued in 1990, 1992, and 1995) that conveyed initially a sense of doom and gloom about the computer and electronics industry. The reports did not speak particularly to government programs, although one of the major objectives of the HPCCI was to keep the U.S. computer industry competitive.

Intellectual Property Issues in Software was published in 1991. A House bill is in committee now on intellectual property issues in electronic media, which indicates that this issue does not go away. It also reaffirms the contribution of CSTB in pointing out that we have patents and we have copyrights, and maybe software has to fit in somewhere either between or beyond them.

Computing the Future: A Broader Agenda for Computer Science and Engineering was released in 1992. We in government thought it was very helpful that the report's first recommendation was to continue to support HPCC. It also spoke to academics, saying that they should broaden academic computer science and engineering. Prophetically, the report recommended broadening HPCC to mission agencies, and this certainly happened. In 1992-1993, several additional agencies signed on to the HPCCI: the National Institute for Standards and Technology (NIST), National Institutes of Health (NIH), National Oceanographic and Atmospheric Administration (NOAA), National Security Agency (NSA), and Environmental Protection Agency (EPA) are examples.

National Collaboratories: Applying Information Technology for Scientific Research (1993)—I think William Wulf can claim a great deal of personal credit for this report. Again, it set a direction that we are now following within agency programs and in the country as a whole. We are trying hard to follow many of the lessons covered in the report.

Information Technology in the Service Society: A Twenty-First Century Lever (1993) bemoaned the problems of productivity and the absence of productivity improvements that could be traced directly to computing advances. We are still struggling with this issue. How do we even count, much less affect?

Then, in 1994, *Realizing the Information Future: The Internet and Beyond* revisited the first report that CSTB had done about the Internet, going back and saying, "Okay, wow, this was really successful. Now what do we do with it?" One of the things this report emphasized was the importance of the concept of the "bearer service" to allow a broad range of applications to be served by a hardware infrastructure.

Research Recommendations to Facilitate Distributed Work, released in 1994, is particularly close to my heart because the Department of Energy commissioned it. It provided DOE with a good research agenda for how to promote telecommuting, which is, of course, becoming more and more ubiquitous. DOE cares because we are worried about energy usage.

Evolving the HPCC Initiative to Support the Nation's Information Infrastructure, the famous Brooks-Sutherland report, was released in January 1995. Out in the anteroom of this meeting is the chart from that report (reprinted here as Figure 2.1) showing how research has affected practice and the long time frame required for some of it. These were helpful lessons for those who thought that the research we do today should be in products tomorrow, and if it is not, it has somehow failed. The report also says, "Do not put all the computer science eggs in the HPCC basket," as well as, "Adjust what you call HPCC so that you keep the books properly."

Also in 1995, *Information Technology for Manufacturing* did an excellent job of laying out a research agenda to address how computing can help on the factory floor. The report is, in a sense, a companion to *Information Technology in the Service Society*; however, it argues that because the shop floor allows better productivity measurement, it is going to be easier to show that computing has actually helped.

Finally, I would say that the work of CSTB past, as I have gone through these reports, has had a substantial influence on the administration, on Congress, and on the research and development community. My little tour through these reports helped me to go back and support that statement by asking: Well, what was really happening in 1986 and 1988 and 1992 and so on? What were the important issues? Did CSTB address them in a productive way? Did its recommendations help?

Paul R. Young

I was very interested in David Nelson's introduction, particularly vis à vis NSF in the past (see Figure 2.2). He pointed out that 10 years ago we were struggling with what to do with supercomputing and networking, establishing the NSFnet. I arrived late this morning because I gave some introductory comments to the panel that is talking about the new supercomputer centers' program, the Partnerships for Advanced Computational Infrastructure. Things from the past continue in revised forms. I think a lot of CSTB's reports continue to influence the future and what we are doing in the present.

I was also amused because David mentioned networking, and we are in the middle of a major change in the NSF networking program as we head to high-end networking. As he pointed out, this program has been influenced by the Kleinrock-Clark report (*Realizing the Information Future: The Internet and Beyond*), which has had a very strong impact not only on policy issues, but also on how we set a research agenda.

Finally, David mentioned that, 10 years ago, we were struggling with the issue of what would become of some initiatives in computing and communications. In 1991, a bill was passed on high-performance computing and communications that goes through 1996 and expires at the end of this fiscal year. The administration and the interagency process struggle with how to redefine this concept and with identifying the future role of federal investments in research in computing, communications, and information.

So in some sense, policies evolve—they do not change. CSTB's reports help us to define these policies and to decide what we are going to do about them. In this context, I was tempted not to discuss the national context in which some of these decisions are made. However, in talking with Anita Jones last night, she suggested some

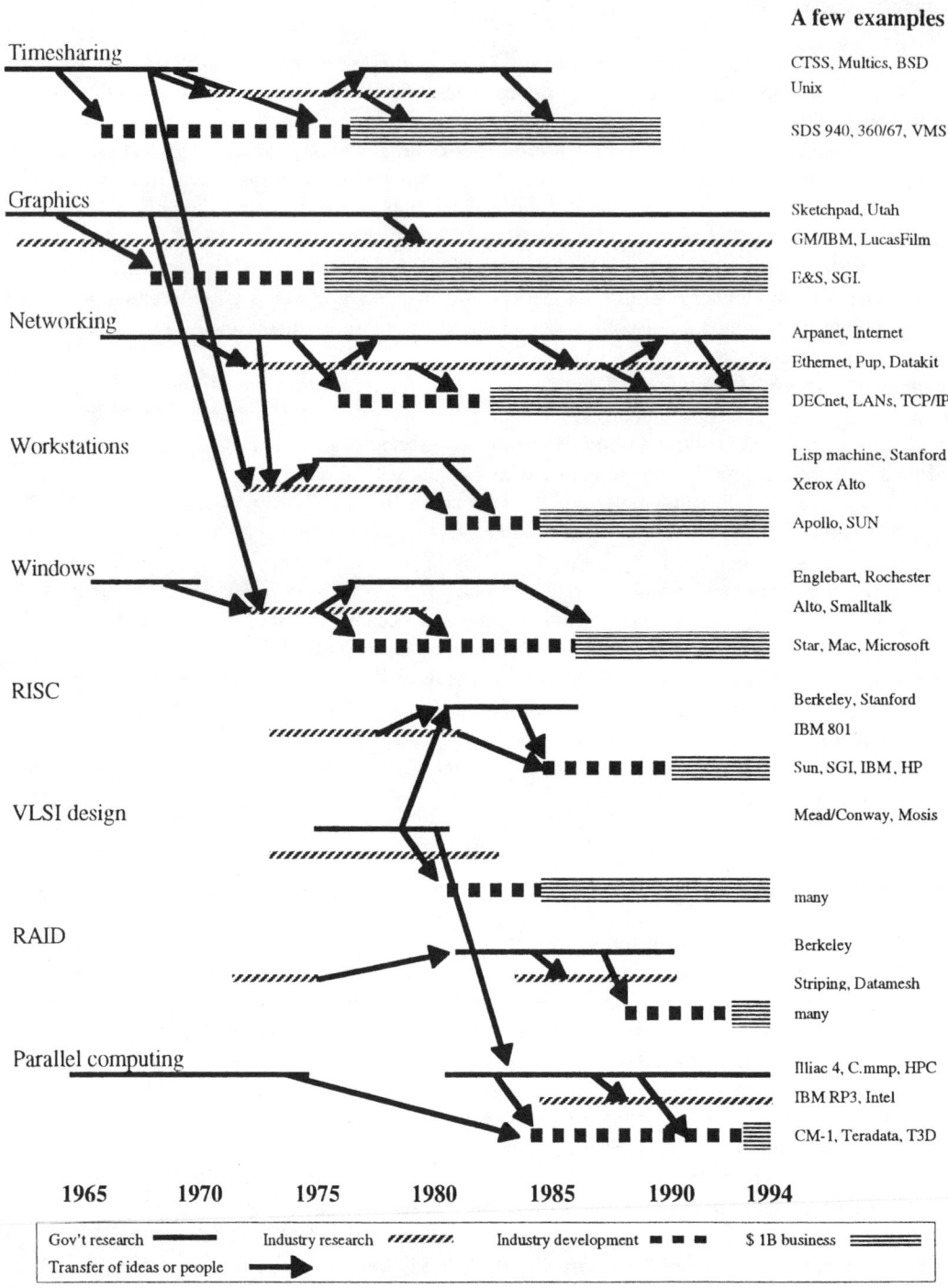

FIGURE 2.1 Government-sponsored computing research and development stimulates creation of innovative ideas and industries. Dates apply to horizontal bars, but not to arrows showing transfer of ideas and people. Reprinted from Computer Science and Telecommunications Board, National Research Council. 1995. *Evolving the High Performance Computing and Communications Initiative to Support the Nation's Information Infrastructure.* National Academy Press, Washington, D.C., Figure 1.2.

PAST	FUTURE

• General-purpose networking for research and education
• High-performance computing
    • Supercomputing
    • Parallel computing
• Grand/National challenges
• Computational science and engineering
• Education and human resources
	• Defining a broad research agenda
	    • Computing systems
	    • Convergence of computing and communications
	    • Human-centered information systems
	• Computational science and engineering
	• Cross-cutting research
	• Education and human resources
	• Learning and intelligent systems
	    • Learning technologies
	• Integrating research and education
	• National infrastructure
	    • High-end computing
	    • Networking
	• Outreach
	    • Partnerships
	    • International

FIGURE 2.2   NSF issues: then and now.

things that might be useful. Michael Nelson spoke in his introduction about his experience with geological time and how things get done in Washington. There are some reasons for this.

On the administration side of how policy is made, there are a lot of players, and NSF often sits in the middle. There is an Office of Science and Technology Policy, a National Science and Technology Council that works through the Committee on Information and Communications, an HPCC process, and a lot of different subareas that define research agendas. On the other side, through the Department of Commerce, there is the Information Infrastructure Task Force that addresses information technology implementation issues. Somehow, the advice that CSTB provides has to influence and pull all of these together into a common framework.

There is a similar method of arriving at priorities in the National Science Foundation that is unique to NSF. You should not think, of course, that CSTB's audience is just those of us who have the immediate responsibility for budgets. There are a lot of people to please: the National Science Board, an advisory committee, workshops that we hold, proposal pressure, and peer review.

Nevertheless, it is clear that external studies are important to us. Peer review is important, and the most prestigious of these reviews come from the Computer Science and Telecommunications Board. Figure 2.3 addresses some issues that I think are important from an NSF standpoint. They provide some indication of places where we might be looking for help from CSTB, without saying exactly how these needs might arise.

First, we continue to redefine the research agenda within NSF, but we do this across the federal government as well. I am sure Howard Frank will say more about this process. CSTB really helps us to formulate this agenda and explain it to the various stakeholders. Some issues that NSF is particularly concerned with include the following: How can we do more interdisciplinary research, and what is the role of computer science and engineering, including computational science and engineering, in that? What is the proper home? What is the relationship between computer science and engineering and computational science and engineering?

Within NSF, we are currently interested in cross-cutting research that addresses human factors. People from different disciplines—biology, cognitive science, and engineering— are addressing questions such as, What does it mean to learn? What does intelligence mean? What is the distributed nature of this? Not all of this discussion bears directly on information technology, but much of it does, and we are trying to identify the relationships.

We are interested in part of the current presidential initiative on learning technologies, a pet project of mine.

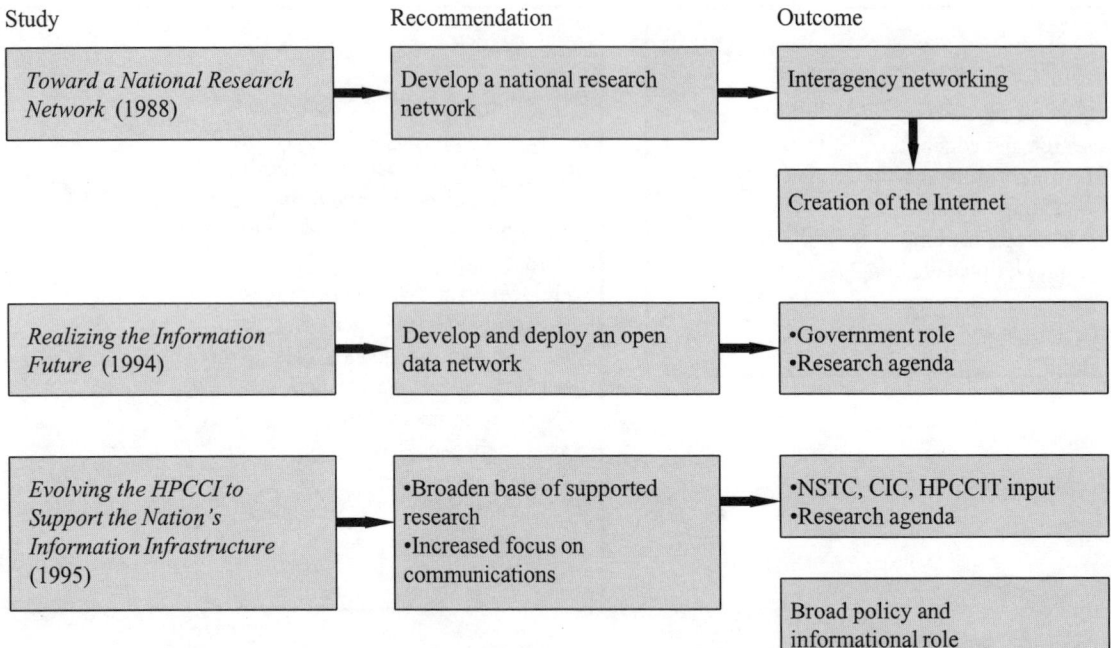

FIGURE 2.3 Illustrative impacts of CSTB studies. NOTE: NSTC = National Science and Technology Council, CIC = Committee on Information and Computing, HPCCIT = High Performance Computing, Communications, and Information Technology (a subcommittee of CIC).

We all know that information technology is going to radically transform education in the twenty-first century, but how? Do we have a road map for this? Let me provide two particular examples. We know that computational methods have transformed how we do much of science and engineering. Computation has joined experimentation and theory as a paradigm for many aspects of how we do science and engineering. When the technology becomes cheap enough and available, it can similarly transform how students learn. It will change the way they think about abstract problems. Do we know what the technical roadblocks are, and do we know what the psychological impacts are? Could CSTB help us address these kinds of questions? I think so.

I would like to come back to collaboratories, an area that William Wulf has pushed and how they can help science and engineering generally. We are going to have some form of collaboratories across geographic boundaries that will influence education. Are these going to be the same technologies? Are children going to use these in the same way as scientists and engineers? What are the technical roadblocks to achieving this by the year 2010? CSTB could help NSF answer this kind of question.

You see Figure 1.2 of the Brooks-Sutherland report again and again (Figure 2.1 in this volume). You know this figure, you love this figure; I know this figure, I love it. It helps to set policy because it shows a very nonlinear picture of the role of federally funded research in the economy and the nation in general. I have found this slide enormously useful. It enables us to talk with policy makers and the general public about the fact that research has a long life. It is still goal directed. It enables us to talk about the fact that the process is nonlinear and that the particular goals may be diffuse in the interaction. This figure has been very effective.

CSTB has another study, "Innovations in Computing and Communications: Lessons from History," currently under way. We are looking forward to the results of this study in particular because the government increasingly is moving toward performance-based budgeting. CSTB can help us with this. The initial idea for performance-based budgeting for research across the federal government was that you were supposed to indicate what the output of a particular program would be in the next two to three years. Of course you could measure the number of publications, but this is not what you are after. You could measure impact. Can you predict this in advance? I do not think so.

We need more help in selling the research mission to the public as a whole, to Congress, and to people down the street. I think CSTB can help us with that.

Howard Frank

I have been involved with CSTB and the earlier telecommunications board for quite a few years. Having been in government, I have been involved in two or three different ways. When Mike asked me to speak about what CSTB should be doing in the future, at first I was taken aback a little because, well, how do I know? Then I decided that the key to understanding what CSTB should be doing is to understand what is happening in information technology in general.

What I will do is give you a picture of what my personal feeling is. This is not DARPA's position or anybody else's, but what I think the challenges of the future are. If you looked at the list of topics that David Nelson has talked about, CSTB has been involved in all of the real issues of the past and present—right on top of them—sometimes a year or two before the fact, sometimes a year or two after. The history of CSTB has tracked the history of information technology.

So I thought I would tell you where I think the problems are. I will point out that there is an anomaly here. We are living through the greatest revolution in information technology that the world has ever seen. The economy is robust. The technology sector is leading the stock market. In the last week there was probably a 5 percent rise in the stock market led by the technology shares, and so on. Nevertheless, I think the future may be rather dim. I think the first challenge is that, in this nation, computer science is basically an obsolete field. Many things that computer scientists have worked on have been wonderful and great. However, some critically important things have been missed.

It was good that we had an introduction that talked about information warfare. When I first became director of the Computing Systems Technology Office in DARPA and began talking to my staff about survivability of the infrastructure, I got the response, "Well, we have already solved that problem. The Internet does adaptive routing." I said, "Wait a second, you do not really know what I mean." A year later, we tried to engage the academic community on the topic. We discovered that there was no academic community engaged in the topic.

This is just one example of the fact that computer scientists have become smug in the wonderfulness of the technology that exists—technology that has been created, by and large, in spite of them rather than because of them. It does not mean that there have not been great things that have come out of the academic community, but the academic community certainly has not led.

Long-term information system technology research in this nation is in great danger. This is an area where we have to cry for help right now. Over the past five years, there has been a collapse of long-term research and development in the industrial sector. Nobody knows exactly how big long-term research really is in the industrial sector, because when companies budget research and development, it may look like $20 billion, but the development part may be $17 billion, $18 billion, or $19 billion.

We know that the Bell Labs of the past is gone. There is tumbleweed blowing down the halls of IBM's research facilities and in many of the other major industrial organizations as well. In the federal government, we are under tremendous pressure—there is vast misunderstanding of the relationship of long-term research and development to the nation's prosperity and future. This is the same problem that CSTB was looking at 10 years ago, except now it is worse. We probably will not know the results of it for another decade or more.

High-end strategic computing is in danger of collapse. We have seen great technological success in terms of the high-performance computing program itself. It introduced the concepts of parallelism and scalability into the commercial world. If you look at medium-scale computing, it now reflects the results of the government program. If you look at the very high end, however, the market is in danger of disappearing. The very high end of strategic computing is a government marketplace, not a commercial marketplace, but there is no government research program that yet recognizes that fact. There is tremendous pressure from Congress and from our own internal constituencies to cut back on what had been considered high-end strategic computing for homogeneous computing systems. We really do not have a strategy for how to continue to acquire computers from a marketplace that cannot afford to spend the R&D dollars necessary for the high-end computation required for many government-unique problems.

Finally, and this is an anomaly, we celebrate the wonderfulness of the Internet when I believe it is now moving into a period of decline in performance. The vast social phenomenon arising from the Internet is yet to take place. We see just the beginnings of this phenomenon. Yet yesterday, David Nelson presented some initial results about performance on the Internet for the research community. We need another study called *Toward a National Research Network* because we no longer have an adequate national research network. The exploding user population has reduced performance below the levels needed by the research community.

What are the implications of all of this? I think CSTB is one of the only places on earth that can deal with this class of problem. It has the only group of people who understand not only the technology issues, but also some of the social and political issues. CSTB needs to become much more proactive. I want to use one specific example. This was not CSTB, but it was the National Research Council (NRC). When I was on the study committee that looked at the survivability of the U.S. telecommunications system, we came to some pretty significant conclusions in 1988, 1989, and 1990. We decided to entitle our report *The Emerging Crisis in the Telecommunications System* because we felt that this is what it was. The report ended up by being called *The Growing Vulnerability of the Public Switched Networks*. That is my point. CSTB needs to become a much more proactive organization that not only helps give insight into policy, but actually helps to beat people on the head until policy is changed.

DISCUSSION

MICHAEL NELSON: Before we go on, I want to add one thing that I forgot to mention. Even though this is the *National* Academy of Sciences and the *National* Research Council, CSTB is having a growing impact internationally. I meet with a lot of people from other countries who want to know how to build up the Internet in their country. They are learning about resources that CSTB has provided, in part because of Marjory Blumenthal's efforts, because CSTB is now up on the Web and because every time I travel overseas I take approximately 30 pounds of your books and hand them out. I used to take only the little red book, *Realizing the Information Future* (1994). Now I have to take that and *The Unpredictable Certainty* (1996). In the past six months, I have handed out dozens of copies in Russia, Beijing, Singapore, and Jakarta. They have been read, and they have influenced a lot of thinking. So I commend you for this effort, and I hope that you will continue to reach out.

EDWARD FEIGENBAUM: CSTB operates in the framework of the commissions and councils, and so on, of the NRC, which is in the framework of the National Academy of Sciences. All are extremely slow and conservative organizations, unwilling to say things that make anyone bristle. So a lot of what CSTB might try to do is either squashed or squashed in advance by this elaborate structure.

I want to point out how long it takes to get a CSTB report out. It takes forever. Bob Lucky has a comment in the published abstracts about the fact that we are zero for one on predicting Webs. The World Wide Web was invented and Mosaic was invented and had 1 million users—all within the time frame of one CSTB report.

MICHAEL NELSON: I would like to second that. Being in the policy-making process, I always want the answer tomorrow, if not yesterday. The only counter to this argument has been that CSTB has been ahead of the curve in identifying the issue that will be hot in two years. This is good, because it takes two years to write the report. I hope the process can move faster. Part of this is our own government's problem. The encryption study, which is coming out soon, has now completed the NRC review process. It would have been nice to have the report six months ago, but it is going to be incredibly timely. In a way, the need for that report is growing, and I think it is going to influence decisions that are just about ready to be made, so the report is right on target. Some other reports have been a little late, but I think the NRC, and especially CSTB because it anticipates, has done better than some other organizations in Washington.

I should add, however, that we are losing some of the other organizations that have supplemented the work of the Board. The congressional Office of Technology Assessment has disappeared. The Annenberg Program from Northwestern, which has done a lot of work on telecommunications policy, has been phased out. So the demand for the products of this Board, and the demands put on it, will be even greater.

Any other reaction to that comment?

PAUL YOUNG: I am very sympathetic to the time issue. It would be very nice to cut the time of producing reports in half. This said, there is a lot of policy made in Washington by the close of business today, and there is definitely a place for a group that can take a deep, measured approach and think things through very carefully. I think that this really has to be preserved while one speeds up the process because it is very easy to shoot from the hip. You see a lot of it.

MICHAEL NELSON: I should also say that there are reports that, although they are toned down a little bit, can still deliver a very powerful message if they are delivered personally by the people who helped write them. The other thing I would like to see more of is taking these reports and really pushing them in the policy process. People who get the reports often do not read them until somebody comes in the room and says, "Here it is. This is why you have to read Chapter five." This, I think, is something Marjory and the team have been doing more frequently. Many of you have been in my office; you have been in a lot of offices. I think the broader community needs to take these messages and get them to policy makers.

MICHAEL DERTOUZOS: I am not sure that I heard what I heard from Howard Frank. Did I hear sort of a Nostradamus thing? I would like to perhaps profoundly disagree or agree because I am not sure I got what was said.

I agree our field has become narrow. I see tremendous opportunities ahead. Some predict there will be 1 billion interconnected machines by 2005. I see 15,000 independent software vendor artifacts going for those machines. I see the entire theory of computer science moving away from the single machine and addressing what happens out there when you have billions of machines. We do not have Turing theories for this. We do not have systems theories for it. We do not have software for it. We do not have systems for it. All this has to happen. We tried doing artificial intelligence things and they did not pan out. This does not mean that they are wrong. It took 250 years to progress from steam to jet power, and in computing we have had only 35 to 40 years.

HOWARD FRANK: Mike, I think you are violently agreeing. The opportunities are there.

DERTOUZOS: But let me have my tantrum. I think there is just a wonderful world ahead, and I am certainly excited about it.

FRANK: There is a fantastic world ahead. It would be nice if we had some of the theory now.

DAVID NELSON: This is the perennial discussion, and it has taken place in mathematics and other fields. My personal view is that it helps to have very good academics working in the field, rubbing shoulders with those who are trying to apply it. In the Department of Energy, we are bringing together mathematics and computer science in jointly funded projects that are trying to do applications. There is, of course, a danger that you may degrade the quality of the research, and I keep my finger on it. I keep asking people who are managing those programs, and the answer I am getting is, no. This is invigorating and stimulating.

WILLIAM WULF: I would like to say just one thing about the issue of the time it takes CSTB to complete reports. I think this is a serious issue, and I keep pushing the staff about it. Part of the problem is internal to the NRC, and that is the part we have the potential to do something about. However, delays frequently occur at the front end. Sometimes CSTB does not get a contract for a long time. This has been a problem on the Ada study, where we are trying to do a fast turnaround.[2] So I am going to put in a plug for why it is so important that agencies like yours have provided sustaining funds—core funds, as they are called—because those sometimes allow CSTB to get a quick start on projects.

---

[2]Computer Science and Telecommunications Board, National Research Council. 1996. *Ada and Beyond: Software Policies for the Department of Defense.* National Academy Press, Washington, D.C.

3

The Global Diffusion of Computing: Issues in Development and Policy

Seymour E. Goodman

THE GLOBAL DIFFUSION OF COMPUTING

Twenty years ago, computing was very much an American domain. A handful of American companies manufactured most of the computers in the world, and annual machine production of the biggest companies measured in the tens of thousands. Most of this equipment stayed in the United States, and little of the rest moved outside of industrial countries, most of which were American allies. Foreign manufacturers (mainly in Western Europe, Japan, and the former Warsaw Pact nations) paled in comparison. The ARPANET had about a hundred hosts, with minimal connectivity with England and Norway, and no comparable network existed outside the United States. Computers were sparsely scattered outside the First and Second World nations.

Today's computing world is profoundly changed. Global production of microprocessors and computers amounts to tens of millions of dollars annually. Capable manufacturers exist in many places around the world, including several nations that were technologically backward in the early days of computing. The United States itself imports a lot of computing equipment, mostly from East Asia and from the foreign manufacturing sites of American companies. International sales of many important American companies amount to about 50 percent of total sales. The ARPANET has evolved into the Internet, which has about 10 million hosts. The non-U.S. share of these hosts exceeded 50 percent for the first time in 1995 (the foreign share of computers in use passed that milestone about 1990). More than 170 countries have some worldwide network connectivity, including Third and Fourth World countries such as Bangladesh, Mongolia, Mozambique, Peru, and Romania. Today, there are about half as many computers on the planet as there are cars, trucks, and buses.

More generally, information technology (IT) has become closely associated with modernization, with being part of the global economy, and—in one form or another—is considered necessary or highly desirable almost everywhere. According to some, IT will help bring the world together, paving the way for economic development and even sweeping aside the traditional borders (and governments) that divide people. (We might also note that the global diffusion of IT is hardly uniform and that, in some ways, IT is causing increased stratification.)

NOTE: L. T. Greenberg also contributed to the content of this presentation.

PLAYERS ON A MUCH EXPANDED FIELD

The United States has had an enormous role in stimulating and enabling this change. American contributions include cheap and powerful hardware; great quantities of (often "free") software; networking; American buying and importing habits (which recently include a trade deficit in IT products); the "Mecca" of the U.S. technical education community; and the fruits of military and space research and competition (most notably, from a global standpoint, the Internet and global positioning system). The United States still comprises about 40 percent of the world's computing market.

Although its relative position has declined, the United States still has almost as much computing as the rest of the world together. This country is now the most important player in a much bigger game. As such, the United States is more prominent, accomplished, and extensively involved in computing, and more people in the rest of the world have direct contact with the United States than ever before. Thus, more foreigners than ever before care about what is done by the United States.

As important as the United States is in global IT, there are also many other players. Some are already quite substantial and others are becoming so. These players are not just governments of countries that now have a serious presence in the global IT scene and are not just foreign IT companies. Rather, the spread and application of IT has been such that many nongovernmental, non-IT industry stakeholders have emerged or have been strengthened. Their involvement results from their interests in the applications of IT (including its "application" as their exports), and they are empowered because of their wealth, constituencies, and newfound voices (often enabled by IT). The interests and the numbers of these actors also create new, or exacerbate old, conflicts and issues. These stakeholders cover a much broader spectrum of business, academic, nongovernmental organization, and private citizen actors than ever. There is even a larger set of stakeholders within governments. Furthermore, the global diffusion of IT has weakened some older stakeholders, and they have not all taken their diminution quietly.

To illustrate how international changes in value conflicts and stakeholders have affected a major technological policy issue over the past 10 years, consider U.S. national security and foreign policy export control. This is arguably the earliest, longest-lived, computer-related U.S. public policy issue, and one with an enormous footprint across the U.S. government. Export control was the subject of one of the first major Computer Science and Telecommunications Board (CSTB) studies,[1] as well as one of its most recently completed major studies. We limit this discussion to export controls on computers, particularly high-performance computers, but a similar short analysis might also be made with respect to the control of encryption.

First and foremost, what has changed in the world is that with the demise of the Soviet Union, the peer threat pursuing a full spectrum of advanced military systems has all but disappeared. This is not to say that there are no significant threats to the United States from foreign sources who may benefit from enhanced computing capabilities, but these threats are not in the same league as the earlier possibilities for global superpower conflict. Taken together with the realities of technological progress and diffusion, this means that advocates of controls find it harder to justify the continuation of controls on the basis of specific applications that potential enemies might be able to pursue using computer products that might still be effectively controlled.

To appreciate how difficult effective control has become, we note that in early 1993, a raw computing performance threshold of 195 MTOPS (millions of theoretical operations per second; since appropriately revised upward twice by the U.S. government) was used to define a control regime for "supercomputers." Today, millions of *microprocessors* are produced each year with performance levels of about half to four times that value. There is now so much made-in-the-U.S.A. computing in so many parts of the world that it is unfortunately inevitable, but essentially unpreventable, that some of it will be used against U.S. national security interests.

Furthermore, there is now a greater division of value judgments within the national security group of stakeholders. For example, some Department of Defense beneficiaries of COTS (commercial, off-the-shelf) high-performance computing (HPC) see more benefit to what they do and to U.S. national security from a strong, technologically vigorous, domestic industry than from continued export controls.

---

[1]Computer Science and Technology Board, National Research Council. 1988. *Global Trends in Computer Technology and Their Impact on Export Control*. National Academy Press, Washington, D.C.

The extent to which foreign governments, even those of our closest and former Coordinating Committee for Multilateral Export Controls (COCOM) allies, consider the value of export controls covering what are increasingly COTS dual-use technologies is not obvious. None gives it the same level of visible attention. When computing was much more exclusively in the U.S. domain and these allies shared a common threat to survival, it was easier for all to agree to controls. In today's world, it is unimaginable how an effective unilateral control regime might be constructed.

Turning now to stakeholders in business and industry, we can find several factors that expand the set of participants. One is that technological progress has been such that HPC, at least by the early 1993 export control definition above, is no longer limited to the extreme high end of the total industry product line. It now covers a substantial fraction of the computers in the world, and thereby many of the computer makers, a significant fraction of which are not on U.S. soil. Furthermore, about half of the new computer sales in the world are not made on U.S. soil.

In many countries, the promotion of business interests is a basic part of national economic policy. It is even fashionable for many countries to be formulating explicit national informatics policies. In the United States, which does not have a comprehensive national industrial policy, there are, nevertheless, many government players promoting business interests at the national, state, and local levels. With many government stakeholders and their constituents, the concern for jobs and exports carries more weight these days than claims of computer-based military threats (although computer-based security problems have been attracting a much greater level of concern in the business and privacy arenas, partly as a result of global diffusion).

Today's world of computer sellers and buyers—including huge multinational corporations (MNCs), many technologically capable small companies, and an immense consumer community ranging from governments to other large MNCs down to millions of small organizations and individuals—constitutes a large and growing, push-pull combination that is under pressure from intensifying global competition. As in the drug trade, there is a huge traffic among willing and resourceful buyers and sellers, and it is difficult for third parties with conflicting interests to interfere with this, for example, through export controls, tariffs, or import prohibitions. Some vast and wealthy industries (most notably the entertainment industry) depend increasingly on IT as a vehicle to do their business worldwide, adding a powerful "pull" element to the "push" interests. As IT diffuses more widely and becomes increasingly pervasive in human activities everywhere, this buyer-seller community gains in importance and influence.

There are other parties with stakes in the international diffusion of computing. Most are recent entrants, and almost all favor the diffusion of computing and related technology transfers. These include various players in developing countries who see computing as necessary for development and as a way to join the rest of the world. Others see IT, in the words of Ithiel de Sola Pool, as the "technologies of freedom," supporting openness, civil and human rights, and democracy; encouraging American values; and making it harder for tyrants to hide abuses. The media also ride on this track, promoting freedom-of-the-press values (and for other reasons, for example, to sell entertainment and to provide better facilities for their reporting).

People in these categories might argue that instead of controlling the infusion of computing into Eastern Europe and the Soviet Union in the late 1980s, the United States should have tried to pour it in. They might further argue that the infusion of IT did more to bring about the decline of overall Soviet power than export controls at that time did to limit its military growth.

The academic community tends to favor transferring more IT. This viewpoint arises in a large number of contexts, including support for the reasons discussed above and for other reasons such as promoting academic freedom, encouraging privacy and security through encryption, participating in efforts to bring useful technologies to colleagues in other countries, and forming virtual scientific communities. American academic stakeholders are also increasingly dependent on foreign student enrollments to keep their departments viable. Foreign shares of graduate student enrollments in science and engineering departments in U.S. universities amount to 30 percent or higher. Simple demographics would indicate that faculty shares may follow suit. Such training constitutes one of the most effective forms of technology transfer, and it is being done on a large scale. For example, it is estimated that on the order of 30,000 to 40,000 Chinese students are studying in the United States. How does one justify a major regime for controlling COTS hardware products under such circumstances? It seems like locking doors in a building when the walls are falling down.

Export control is but one example of a major technology issue that has been dramatically affected by the global diffusion of computing and the attendant expansion of stakeholders. The spread of global networking in various forms is creating or greatly exacerbating a number of issues that are attracting, or will attract, further attention. For example, many things from American political correctness to law enforcement practices to intellectual property rights to Islamic religious tenets may be seen as local sensitivities that are increasingly to be battered in a global multimedia free-for-all. Many forms of conflict, both civil and military, are finding fertile international or transnational breeding grounds in the IT-based media.

Many will not like the redistribution of—or compromise of, or assault on—moral or legal sensitivities, jobs, markets, wealth, influence, military power, and governmental authority via the international use of IT. Many of these will be influential stakeholders who will seek help in various ways from their governments or international organizations. There are notable cases in which this has already happened. IT may also enable them, as well as those they oppose, to band together to seek help or assert influence independent of their governments.

CSTB studies, and those of the National Research Council more generally, essentially have tried to bring science- and technology-related trends and issues to the attention of interested constituencies, and the attendant analyses have tried to inform or guide choices for policy formulation. To a large extent, the latter is a matter of balancing or choosing between value conflicts among these constituencies. The primary funders, and the principal targets as consumers of these studies, have been parts of the U.S. government.

The activity levels and pressures on national policy makers in these matters are likely to increase, as will some frustration levels. It appears possible that the relative power of national governments in these matters will diminish, but certainly not disappear, and that governments will have to contend with and work with other players as never before. The activity and frustration levels of advisers to governments, such as the CSTB, may also increase substantially.

The main purpose of this short presentation has been to draw attention to the greater internationalization of both the issues and the affected stakeholders that is accompanying the global diffusion of IT. This is not a short-term or transient phenomenon; it is going to be forever.

4

Engines of Progress: Semiconductor Technology Trends and Issues

William J. Spencer
Charles L. Seitz

William J. Spencer

Everyone knows the adage (variously attributed to baseball managers and ancient Chinese philosophers): "It is difficult to predict, especially the future." With that in mind, this paper focuses on how productive the semiconductor industry has been over the past 50 years, its status in 1995, the road map for the industry for the next 15 years, some of the challenges that the industry faces, a few predictions, and a final caveat.

The productivity of silicon technology is usually measured in the cost reduction of memory or the increase in processing power. The cost of memory will have fallen by roughly five orders of magnitude from 1972, when the first 1,000-bit static random access memory chip (SRAM) was available, to the year 2010, when the 64-gigabit dynamic random access memory chip (DRAM) will be available. The cost of computing has also decreased dramatically during this time frame. *Fortune*, the impeccable source of reliable information, recently showed that the cost of MIPS has declined by a factor of 200,000 over 30 years. This dramatic reduction in cost or increase in computing power over a period of several decades is unparalleled in any other industry. It is this continual improvement in productivity in silicon integrated circuits, and the related lower cost of memory, computing power, and communications bandwidth, that is leading the world into the Information Age.

The productivity gains are made possible by continued technology improvements in the manufacture of silicon integrated circuits. The silicon transistor has been the principal element in this technology since 1957. The types of transistors have changed from grown junction to planar, from bipolar to metal oxide semiconductor (MOS), but the transistor effect has been the principal element in all silicon integrated circuits for the past 40 years and probably will continue to be for at least another quarter of a century.

The technology changes that have occurred are listed in Table 4.1. These begin with the invention of the transistor at Bell Labs-Western Electric in 1947 and continue through the introduction of ion implantation, reactive ion etching, optical steppers for lithography, e-beam mask making, and a variety of other technologies. Except for some of the design and simulation packages, these technology innovations have all come from industrial labs. Some of these laboratories do not exist today; others have been downsized and redirected. This represents a major challenge for our industry that I want to come back to later in this presentation.

In 1995, most leading-edge semiconductor manufacturers were using a 0.35-micron technology for manufacturing MOS devices in microprocessors, memory, and logic. A cross-section of a CMOS transistor pair in this technology is shown in Figure 4.1. Silicon chips today contain as many as 20 million of these transistors. This

TABLE 4.1 Major Semiconductor Innovations

Innovation	Laboratory	Year
Point contact transistor	Bell Labs-Western Electric	1947
Single-crystal growing	Western Electric	1950
Zone refining	Western Electric	1950
Grown junction transistor	Western Electric	1951
Silicon junction transistor	Texas Instruments	1954
Oxide masking and diffusion	Western Electric	1955
Planar transistor and process	Fairchild	1960
Integrated circuit	Texas Instruments, Fairchild	1961
Gunn diode	IBM	1963
Ion implantation		
Plasma processing		
E-beam technology		

SOURCE: John Tilton, Brookings Institution, Washington, D.C.

technology has several interesting characteristics. The interconnect is still principally aluminum, although it is now alloyed with copper and titanium. The interlayer dielectric is still silicon dioxide. You will note that this plasma-enhanced chemical vapor deposition of silicon dioxide leaves very uneven layers. There is a complex metallurgy in the vias that consists of titanium and titanium nitride with tungsten as the principal conductor.

Today, the semiconductor industry is roughly a $200 billion-per-year industry; $150 billion of this is in the sale of semiconductor devices, roughly $30 billion in the sale of processing equipment, and approximately $20 billion in the sale of manufacturing materials, including silicon, mask blanks, photoresists, and production gases and liquids (Table 4.2). In the 1990s, the industry has grown at an extremely rapid rate, averaging more than 30 percent per year over the past three years. The industry is projected to grow at an average rate of about 20 percent per year for the next 15 years.

If we compare the growth rate of the semiconductor industry (about 20 percent per year) with the gross domestic product (about 2 percent per year), by the year 2019 semiconductor sales will be equal to the U.S. gross domestic product of $11 trillion. This leads me to prediction 1: *The increase in semiconductor sales will flatten before the year 2019.*

The U.S. semiconductor industry has developed a road-mapping process that looks at technology needs for the industry 15 years into the future if costs are to continue to decline. An example of the content of this road map is shown in Table 4.3. The road map considers memory, high-volume logic, and low-volume logic. The year represents the first shipment of a product with the given technology. A major consideration of the road map is cost. The cost per transistor for each of these products is predicted to decline with each generation of technology. This leads to prediction 2: *There is no physical barrier to the transistor effect in silicon being the principal element in the semiconductor industry to the year 2010.*

Although the road map projects technology generations for 15 years in the future and the technology required for that projection, it does not define solutions for all of the technology requirements. Let us look at some of the challenges that must be addressed to meet the road map projections to the year 2010.

Will it be possible to design 100 million transistor logic circuits and 70 to 75 billion transistors on memory chips? Figure 4.2 describes what the semiconductor community has called the design productivity crisis. The upper curve shows the compound complexity growth rate in silicon integrated circuits at roughly 60 percent per year. The lower curve shows the compound design productivity growth at about 20 percent per year. This leaves a considerable gap in 1995 and a growing gap into the twenty-first century, which is the consensus of a group of design experts in industry, government, and universities. There is no complete solution to this design dilemma. There is cooperative work on building an open infrastructure that will allow interoperability of commercially available design tools and of those tools that are developed as competitive capabilities in integrated circuit

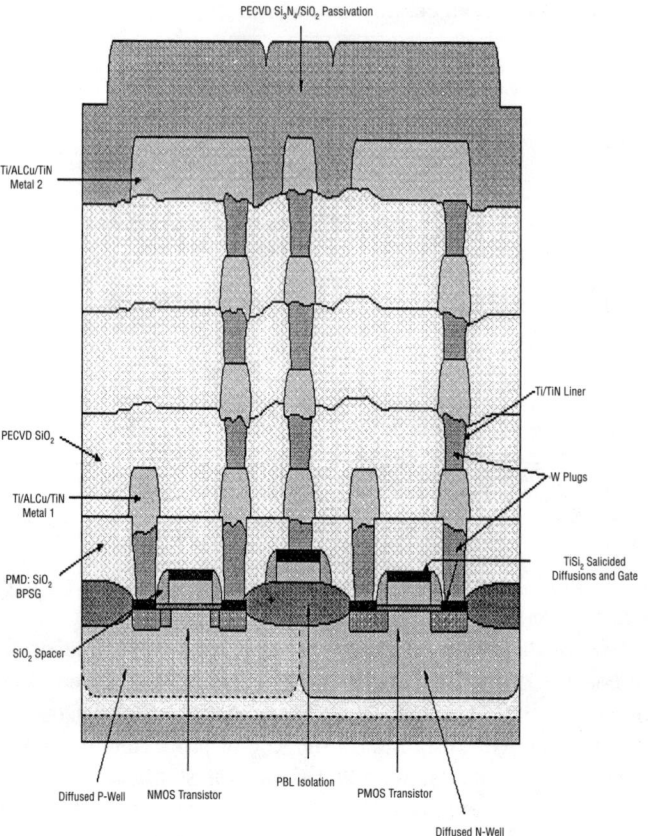

FIGURE 4.1 Device cross section, SEMATECH's 0.35-μm CMOS process. SOURCE: SEMATECH.

manufacturers. The design costs, the size of design teams, and the time required to design products at the end of this century will be major barriers to continued growth of the integrated circuit industry. This uncertainty in design costs and investment to manufacture new integrated circuits has led to numerous joint ventures focused on new products.

Supposing we are able to design integrated circuits in the twenty-first century, will we be able to manufacture them? Here, I would like to focus on just one processing step, lithography. Lithography and interconnect represent the two major costs in the manufacture of integrated circuits. Figure 4.3 shows the road map for lithography and some potential solutions for lithographic requirements. At 0.35 micron and 0.25 micron, the technology choice has been made. Deep ultraviolet (DUV) exposure tools, operating at 248-nm wavelength, will

TABLE 4.2 Semiconductor Industry—1995

Sales	Billion Dollars
Semiconductors	~ 150
Equipment	~ 30
Materials	~ 20
Total	~ 200

TABLE 4.3 Overall Roadmap Technology Characteristics—Major Markets

Year of First DRAM Shipment	1995	1998	2001	2004	2007	2010	Driver
Minimum Feature Size (μm)	0.35	0.25	0.18	0.13	0.10	0.07	
Memory							D
Bits per chip (DRAM/flash)	64 million	256 million	1 billion	4 billion	16 billion	64 billion	
Cost per bit @ volume (millicents)	0.017	0.007	0.003	0.001	0.0005	0.0002	
Logic (high-volume: microprocessor)							L (μP)
Logic transistors per cm^2 (packed)	4 million	7 million	13 million	25 million	50 million	90 million	
Bits per cm^2 (cache SRAM)	2 million	6 million	20 million	50 million	100 million	300 million	
Cost per transistor @ volume (millicents)	1	0.5	0.2	0.1	0.05	0.02	
Logic (low volume: ASIC)							L (A)
Transistors per cm^2 (auto layout)	2 million	4 million	7 million	12 million	25 million	40 million	
Nonrecurring engineering cost per transistor (millicents)	0.3	0.1	0.05	0.03	0.02	0.01	

be required for these two generations of integrated circuits. At 0.25 micron, there is a possibility of additional optical enhancement (OE) through off-axis illumination, phase shift masks, and other optical tricks. At 0.18-micron technology, the solutions are less clear, and for technologies of 0.13 micron or less, it is even less clear which technology will be suitable for production of these devices.

The development of this technology does not come for free. Projected development costs for lithographic systems (including exposure tools, resists, metrology, and masks) at 193-nm exposure wavelengths total nearly $350 million for the 1995-2001 period (costs are highest in the first four years and peak in 1996-1997). This cost is beyond the capability of any single company. This has been an area in which the U.S. semiconductor industry and the supplier industry have worked cooperatively to develop 193-nm lithography and possible optical extensions down to 13 nm. Multicompany consortia have focused successfully on technology development, whereas it is principally bilateral joint ventures that focus on new products.

Now, if we can design (still uncertain) and fabricate (possible) integrated circuits with 100 million transistors, will it be possible to package them? In 1995, 10 million transistors were being built on a single chip with a cost per transistor of about 10 microcents for a total chip cost of roughly $10. The packaging costs in 1995 represent somewhere between $5 and $35 for packages with up to 500 pins. This supports personal computers that cost in the range of $2,000 and provide moderate data capability, no voice, and very slow video. Six years from now, there will be nearly 50 million transistors on a single chip at a cost roughly five times less, with a chip cost that is still about $10. The packaging costs will rise to nearly $50 for packages with up to 1,000 pins. The question is whether this technology will provide components at a cost low enough to keep personal computer prices at roughly $2,000 with better data capability, limited voice, and faster video. The challenge of programs in the government and at SEMATECH is to bring packaging costs down so that they remain on the same level as chip costs. This will require a reduction of roughly two times the package cost.

We have looked at the challenges in design, processing, and packaging of integrated circuits during the first decade of the twenty-first century. What will the cross section of one of these integrated circuits look like? The cross section shown in Figure 4.4 is of a 0.10-micron CMOS process for high-performance logic. There are several major differences between this and the 0.35-micron schematic shown earlier. Notice that the interlayer dielectrics are all flat. This means that the dielectric material has been planarized after it has been deposited. The material has changed from silicon dioxide to a low dielectric constant polyimide. Metallization has gone to copper, with copper plugs in the vias and copper in the liners. All of these new technologies will require extensive design and testing to ensure continued reliability and cost reduction.

Typically, the interdielectric layers are on the order of 1 micron in thickness, while the gate lengths are a tenth

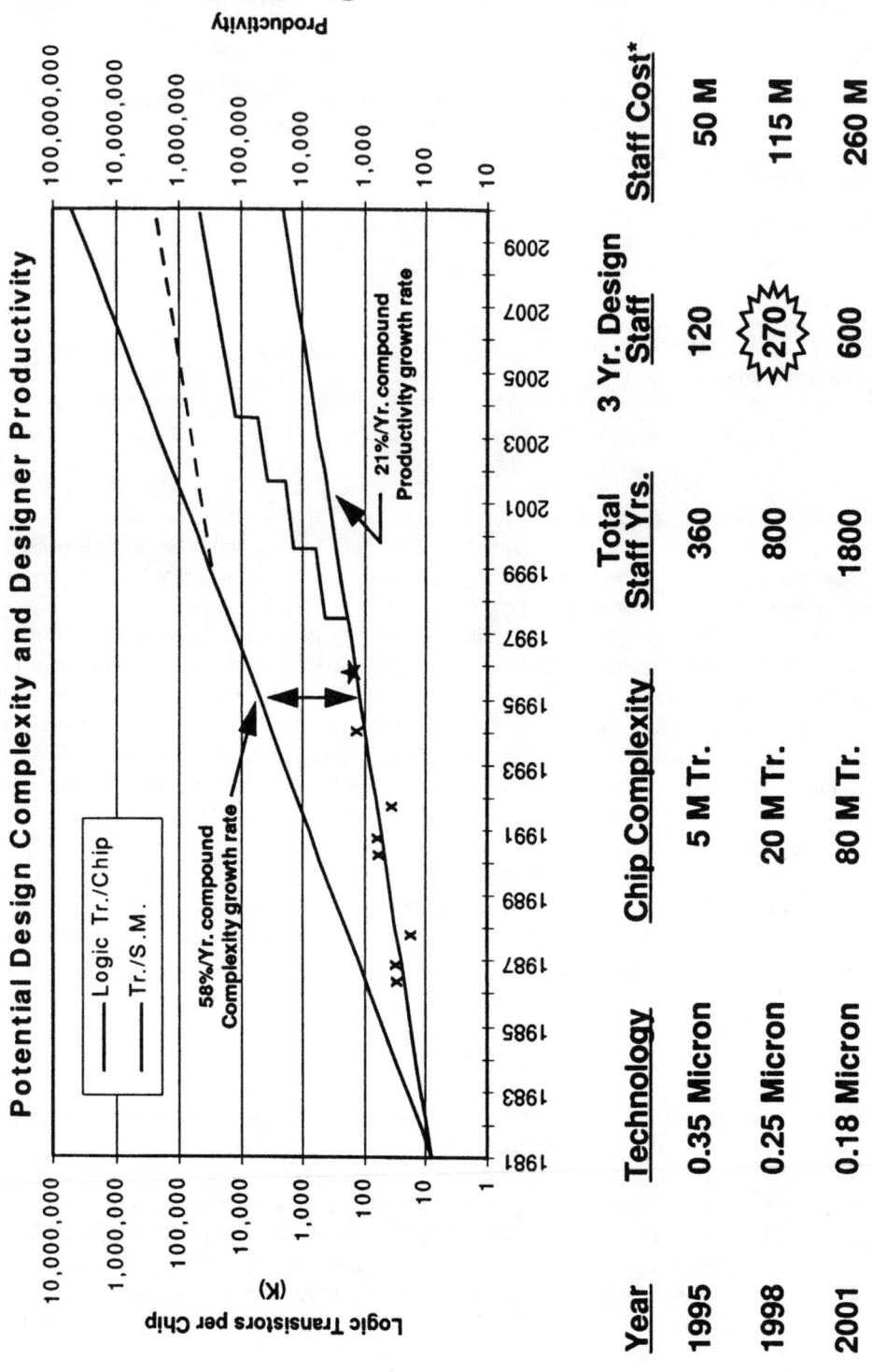

FIGURE 4.2 Design productivity crisis. SOURCE: GWL ETAB Strategic Review, March 1996.

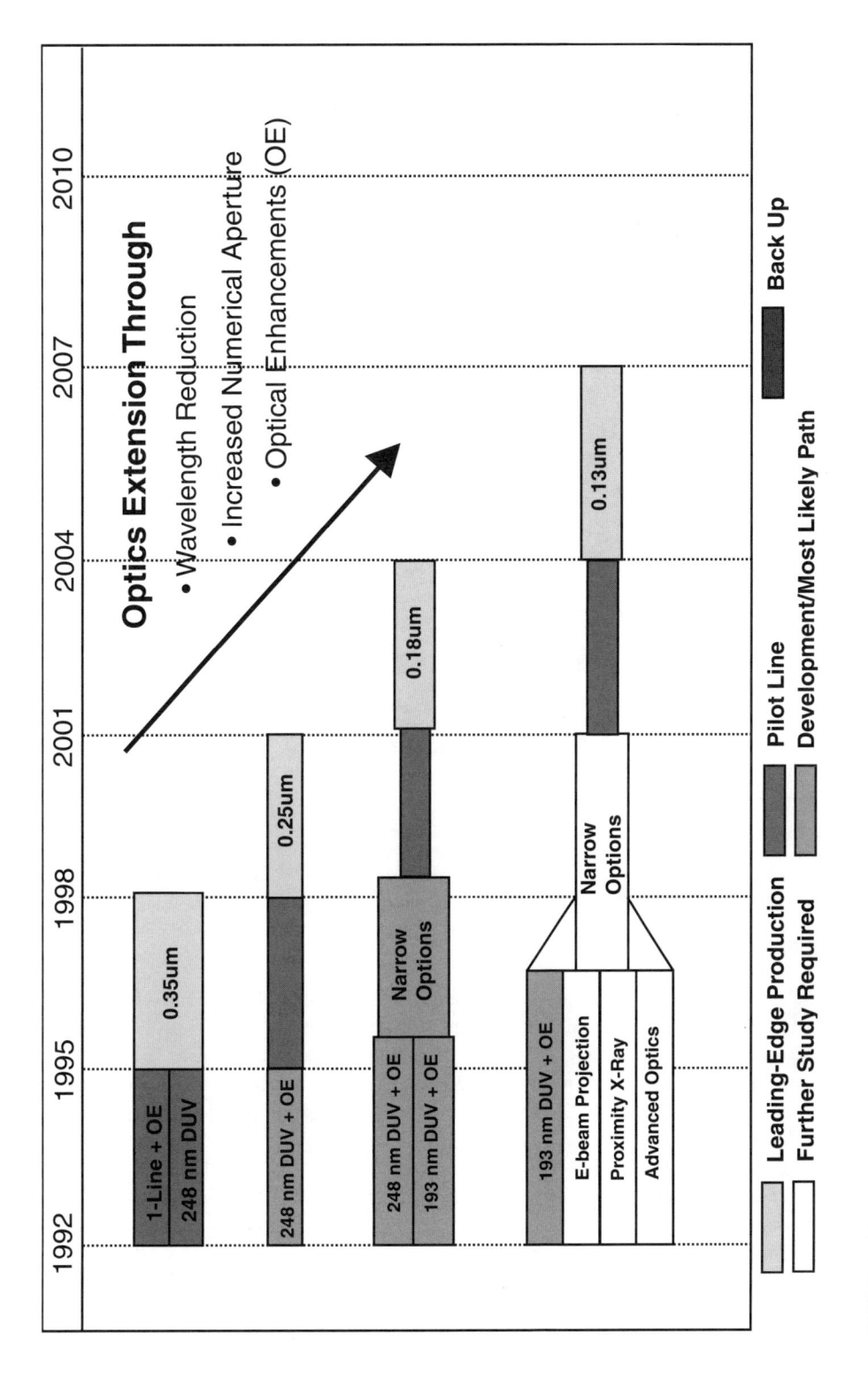

FIGURE 4.3 Lithography roadmap—potential solutions. SOURCE: SEMATECH.

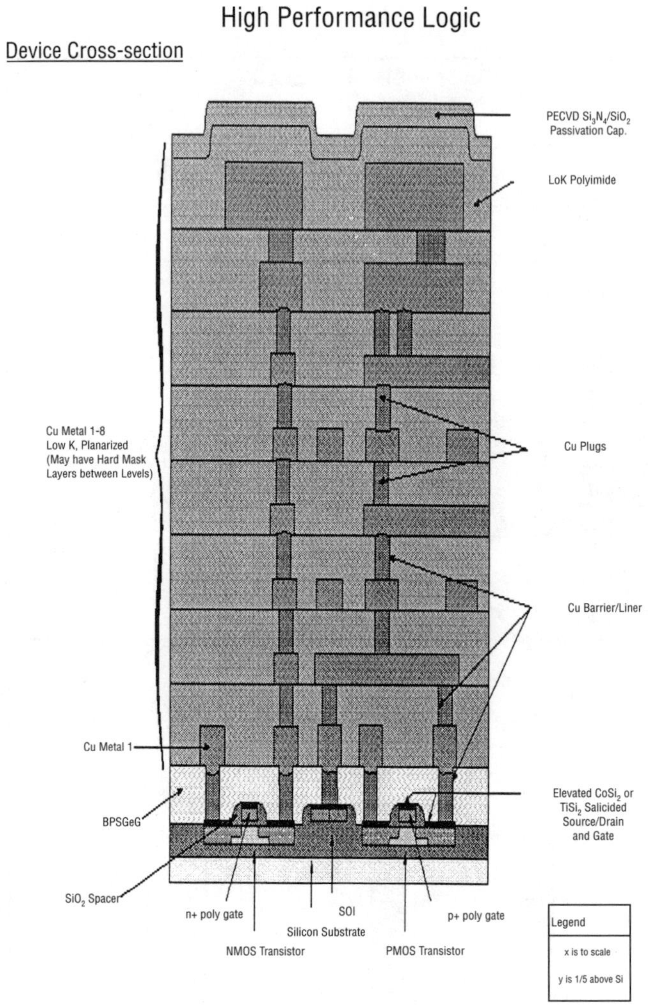

FIGURE 4.4 Device cross section, 0.1-μm CMOS process schematic. SOURCE: SEMATECH.

of a micron. Thus, the schematic in this figure is compressed in the vertical direction and highly expanded in the horizontal direction. If we look at the actual cross section of this 0.10-micron CMOS process it looks more like a skyscraper. This shows why the major complexity in future integrated circuits will be focused on interconnects and the lithography to produce these interconnects. The basic transistor structure, while becoming more difficult to manufacture, will be a much smaller part of the total processing costs compared with the interconnect.

Finally, let us look at the total cost of building a manufacturing facility for future integrated circuits. The costs in 2000 are expected to be about $2 billion for each new fab. In 1970, Intel built its first fab for less than $5 million. The cost of fabrication facilities is growing faster than the growth of semiconductor revenue. This leads to my third prediction: *The entire world's requirements for silicon integrated circuits will be built in a single fabrication facility (probably Korean) by the year 2050.*

Let me go back to a point made earlier in looking at technology changes that have led to the continued productivity growth of integrated circuits and where those technology changes originated. Most of the technology used in the manufacture of modern semiconductors came from industrial laboratories such as Bell Labs, IBM, Texas Instruments, Fairchild, and Phillips. Today, these laboratories either no longer exist or have been significantly downsized, particularly in physical science research. The remaining research is focused on corporate needs, not industry needs, leaving a major technology gap. Without these engines of innovation, the entire semiconductor industry will stall in the first decade of the twenty-first century, and this stalling will slow the growth of computers and telecommunications. There is certainly enough technology to keep the industry going over the next 5 and perhaps 10 years. Beyond this time, there will be serious shortages in design, process, packaging, and other technologies that support the semiconductor industry. This is an issue that must be addressed by the industry, the government, and the universities. There is no simple solution. The semiconductor industry is looking at tripling its investment in university research. Even with this investment, there must be changes in commercialization, culture, management, and education to provide the future source of innovation for the semiconductor industry. This leads to my final prediction: *All of the earlier predictions are wrong!*

As soon as a road map is written down, it becomes outdated. This industry moves so rapidly that it is impossible in any document, talk, or paper to predict where it is going. Usually, the industry has moved more quickly than predictions. Engineers and scientists have typically found ways to overcome technology barriers, the industry has continued to grow at an amazing rate, and productivity continues to increase at 25 to 30 percent per year in silicon technology. This has led to the semiconductor industry as the economic driving force for the Information Age. If it stalls, the applications that are dependent on silicon technology will stall as well. This touches every aspect of our lives, from education to work to leisure. Silicon technology is pervasive in all our lives. You probably have a half dozen or more integrated circuits on your person today—in your watch, beeper, cell phone, and electronic notebook—and if you carry a laptop, of course the number goes up dramatically. There are answers to most of the manufacturing productivity issues for semiconductor technology; the major question will be where our future innovations originate.

Charles Seitz

A remarkable thought—continued progress through at least 2010. The progress described by William Spencer reminds me of neighbors who ask, "Should I buy the new model XYZ computer that just came out, or should I wait another year because they keep getting better and better?" The only answer I have for that question is, "It depends on whether you want a computer to use over the next year."

It has become part of everyone's expectations that computing devices will keep getting better and better. Bill and I are struck by the extent to which the real engine of progress is found in some of the lower levels of computing technology, even though the effect of these steady improvements is to be able to support larger and larger software packages that do more and more.

I am going to take you straight to a designer's eye view of a chip. If you were to walk around the design laboratories at places like Intel and Motorola, or at any of thousands of small companies, you would see people sitting in front of computer screens in cubicles, much like the Dilbert cartoons, with pictures or portions of a chips layout in front of them.

These pictures, like most arcane art, may be a little difficult to interpret. What the designer is looking at is a number of different layers, each in a different color, as if from a skyscraper looking down at a city. Differently colored layers represent layers of metal on a chip. These metal layers are entirely for interconnection. Special symbols, such as boxes with stipple or cross patterns, are used to indicate the connection from one of the metal layers down to another or from the metal layers down to the deeper layers. It would not surprise you, perhaps, that the wider metal layers carry power to the inner circuits of the chip. "Where are the transistors?" you ask. Well, they are in the smallest features hidden way down at the lowest levels of the chip.

To the designer, the transistor is formed where one of the wires—called poly because they were once polycrystalline silicon, although they are now typically composed of tungsten silicide—crosses another wire, the diffused area, down in the very, very tiny and deepest parts of the chip.

The layout of all of these geometrical shapes creates a data file from which optical masks (or reticles) are made, from which chips are made in turn. In this sense, all of the wonderful work that is done in semiconductor manufacturing can be thought of as similar to what the photography industry does. Designers can produce any pattern they want and turn it over to a fabrication facility that can make chips with the same pattern. Although processes for certain types of chips are somewhat specialized, for the most part the fabricator does not use a different process for one kind of chip or another, any more than a printer might use a different kind of paper for a photograph of your daughter or your boat.

"In actual fact," as they say in England, this picture [displayed at the symposium] is not an entire chip. The plot is of an area about a quarter of a millimeter on a side. One of the aspects of chip design that I want to emphasize is complexity, and I will use every opportunity to do so. Remember, Cato the Elder kept repeating to the Roman Senate that Carthage must be destroyed. If you keep repeating something, it oftentimes happens.

So, please excuse the repetition, but these chips are awesomely complex. A state-of-the-art chip today is about 2 cm on a side. Thus, at the scale of this plot, it would take 6,400 of these pages to display the geometry of the entire chip. It is difficult for the designers even to find their way around such chips, let alone to design them.

One of the beautiful things about semiconductors is that they are just physics at the lowest level. You might say, "Why do Bill Spencer and SEMATECH go to such great effort and expense to figure out how to reduce the feature size a little bit more?" Let me try to explain, based on the tabulation along the left side of Figure 4.5.

Starting in the early 1970s, device physicists recognized an ideal form of scaling of the so-called MOSFET (metal oxide semiconductor, field-effect transistor) technologies, in which transistors are formed longitudinally along the surface of the chip. Here is the basic story.

For a scaling factor a, reduce the feature size x to some value x/a. If you do this, you should also reduce the voltages in order to keep the electric fields from increasing; otherwise, you are asking someone to invent new materials that could withstand higher electric fields. (Although chips operate from relatively low voltages, the dimensions are also very small, resulting in electric fields that approach the limits of the dielectric strength of the glass (silicon dioxide) insulator under the MOSFET transistor gate.)

With electric fields being constant in this scaling, and with the mobility of silicon being constant, the velocity of mobile carriers—electrons or holes—is more or less constant. However, the carriers can traverse the smaller transistor in less time, making the smaller transistors (and circuits, if you go through the entire analysis) faster. Every child knows that smaller things are faster.

The reduction in feature size increases the circuit density quadratically with the scaling factor, but if the current is also reduced in proportion, the power per device is reduced quadratically with the scaling factor. It is fortunate that these effects balance, so that the power per unit area remains constant in scaling.

The switching energy, also known as the power-delay product, is a technology metric that closely predicts the cost of a computation implemented in a given technology. The bottom line is that the switching energy, the product of the power per device and the transit time, scales as the third power of the scaling factor. Thus, for example, all of the effort of reducing the feature size by a small factor such as 1.26 pays off by reducing the switching energy by a factor of 2, and this factor of 2 can be applied across the board in all kinds of computing and communications devices. It allows chip and computer designers to offer about twice as much computing at the same cost, or the same amount of computing at half the cost.

Higher circuit density together with larger chips has led to the remarkable complexity scaling of microelectronics shown in the upper right of Figure 4.5. This escalation of the circuits that can be put on a single chip, known as "Moore's law" (after Gordon Moore, chairman of Intel), provides a smooth entry for yet another lesson in complexity appreciation.

One nice analogy is to compare a chip with a city. The minimum spacing of the wires on a chip today is about a micron, whereas city blocks are spaced about eight per mile, or five per kilometer. The difference between the spacing of wires on a chip and the spacing of blocks in a city is a factor of about 200 million.

Let us take one of today's chips, 20 mm on a side, and print a map on it. The multiple layers of wiring on the chip are of greater complexity than the generally single layer of roads found in a city, but you will see that the chip can accommodate the map of a city 4,000 km on the side.

This would be quite a city. If you use the figures from Bill's charts of year 2007 technology, the wire spacing will then be reduced to about a quarter of a micron, raising the scale factor to 800 million. How big a chip can you make? That is determined largely by the defect density. Today, we have defect densities of somewhat less than one per square centimeter, so it is reasonable to build chips that are a centimeter or two on the side. If the defect density is reduced still further, you can achieve acceptable yields on chips that are even larger. A chip 50 mm on

Ideal (MOSFET) Scaling

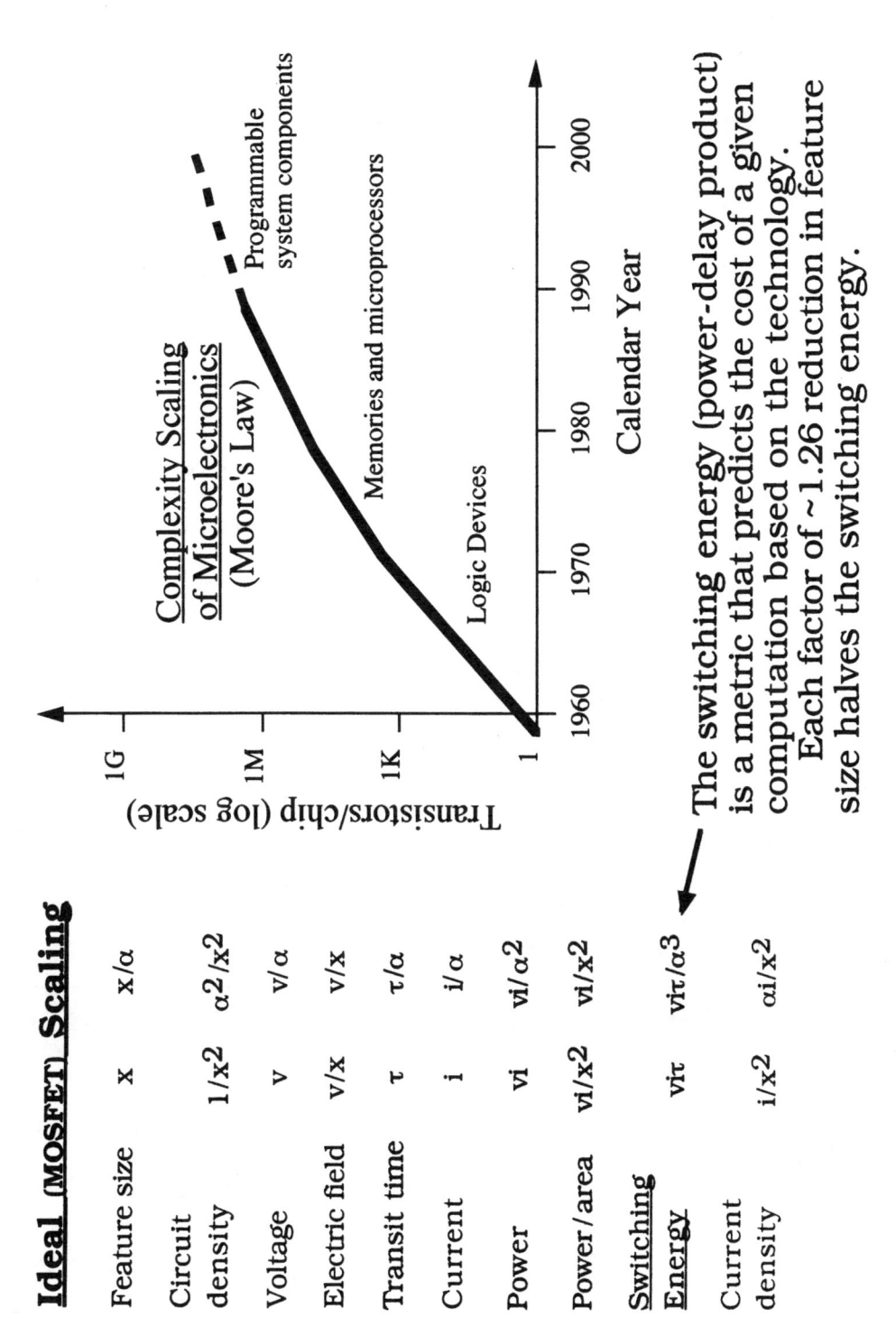

Feature size	x	x/α
Circuit density	$1/x^2$	α^2/x^2
Voltage	v	v/α
Electric field	v/x	v/x
Transit time	τ	τ/α
Current	i	i/α
Power	vi	vi/α^2
Power/area	vi/x^2	vi/x^2
Switching Energy	$vi\tau$	$vi\tau/\alpha^3$
Current density	i/x^2	$\alpha i/x^2$

Complexity Scaling of Microelectronics (Moore's Law)

Transistors/chip (log scale) vs. Calendar Year (1960–2000)

- Logic Devices
- Memories and microprocessors
- Programmable system components

The switching energy (power-delay product) is a metric that predicts the cost of a given computation based on the technology. Each factor of ~1.26 reduction in feature size halves the switching energy.

FIGURE 4.5 The underlying "engine of progress." SOURCE: Myricom, Inc.

> **BOX 4.1 How Do Chip Designers Cope?**
>
> *The "mechanics" of chip design*
>
> Each generation of chips is designed using *computer-aided design and analysis tools* that execute on the previous generation of computers, an example of technology "feeding on itself."
>
> *Respect complexity.* Just as computer software hit complexity barriers demanding the introduction of "structured programming" and other design disciplines, chip designers have adopted analogous disciplines. "Don't design what the simulator can't simulate!"
>
> Submicron devices are not as ideal as the transistors and wires at larger feature sizes, thus additionally complicating the physical design of chips today. In the future, designers may need to learn to cope with the statistical likelihood of a small fraction of the transistors on a chip not working.

a side is what you get if you crank out the math based on Bill's defect-density projections. Applied to the city analogy, the complexity of a single chip would then correspond, roughly speaking, to the area of the earth covered at urban densities at about 10 different levels, including the oceans. Talk about an ecological disaster—the "Los Angelesization" of the world!

Chip designs are, of course, done with the help of computer-aided design and analysis tools (see Box 4.1). The analysis tools provide reasonable assurance that the first time you fabricate a chip, it works well enough for you to figure out what you overlooked. There is, by the way, no easy way to probe the signals in the interior of these chips. Even if you could position a probe on the right point, the energies involved are so small—measured in femtojoules—that the probe would disturb the operation of the circuit.

Designers of processor chips always wish they could use the chip that they are designing to run their analysis tools, but we have to run the design and analysis tools for each generation of chips using computers built from the previous generation of chips. Nevertheless, the technology is improving on exactly the same curve as the demand on the tools, an interesting example of technology feeding on itself.

There was a fairly large brouhaha in the computer software field in the 1970s, triggered by Edsger W. Dijkstra's 1968 article "Got to Statement Considered Harmful."[1] Computer software was getting complicated enough that people had to adopt complexity management schemes cast into disciplines such as structured programming for writing software.

During the past decade or so, we have seen chip designers adopt analogous disciplines. Frequently, these voluntary restrictions are tied into the design tools, just as programming disciplines are frequently incorporated into programming notations. My favorite rule is, "Do not design what the simulator cannot simulate." It may be perfectly possible to lay down some metal and poly and diffusion to produce a certain circuit that would work. However, if the simulator cannot divine that it would work, you had better not use this circuit because you would have to treat it as too much of a special case.

One fly trying to get into the ointment is that, as these devices get smaller and smaller, they are less ideal. Our present devices are what physicists think of as thermodynamic. Their operation depends on aggregates of tens of thousands of charges. The statistical fluctuations around 10,000 are in the sub-1 percent range.

As everything gets scaled down, you reach a regime in which, for example, the threshold voltage of a transistor, instead of being determined by something on the order of 10,000 impurity ions under the gate today, is determined by 100 or so. The statistical fluctuations around 100 may be 10 or 20. In addition, instead of having a mere 10^8 devices on the chip, we may have 10^9 or 10^{10}.

---

[1] Dijkstra, E.W. 1968. "Go to Statement Considered Harmful," *Communications of the ACM*, March, pp. 147-148.

> **BOX 4.2 How Do Designers Innovate?**
>
> Ivan Sutherland's (~1978) story about the bridge builders who knew all too well how to build bridges of stone: when the new material, steel, came along, they cast the steel into blocks from which they made arch bridges. How would the insights come about that would result in the truss-and-trestle bridge, let alone the suspension bridge? In part, from recognizing how the properties of steel differ from those of stone.
>
> How does microelectronics differ from earlier digital technologies? It is
>
> - highly universal (no escape into specialized technologies);
> - severely communication limited—wires use most of the area and power or cause much of the delay (favors simplicity and concurrency); and
> - easier to reduce cost than to increase speed (favors concurrency).
>
> The "success" stories in circuit, logic, and architectural innovation—the dynamic RAM, programmed logic arrays, RISCs and cache memories, highly concurrent computers, and many other innovations—can be traced directly to insights that respected the limitations or exploited the capabilities of the medium.
>
> ---
>
> RAM = random access memory; RISC = reduced instruction set computing.

One basic change in microelectronics technology to be expected is that chips will be made with no expectation that all of the devices on the chip will work correctly. People doing the design and engineering have to start thinking in terms of systems such as the national power grid, in which one expects that some parts of it, at any given moment, will not be working correctly.

Finally, the question that most interests me is, How do designers innovate? (see Box 4.2). I wish Ivan Sutherland (a former member of the Computer Science and Telecommunications Board) were here. I heard a story from Ivan at the California Institute of Technology in about 1978 that I have carried in my head since and thought about a lot. This is a story about bridge builders who build bridges out of stone. Then steel, the exciting, new material, comes along. The builders want to be modern and innovative, so they start to use steel. They use it by casting the steel into blocks, which they then assemble into arches to make bridges.

With new technologies—and microelectronics is fundamentally quite a new technology—there is the same question about its real properties. In its early days, microelectronics was used as a substitute technology for transistor logic. It came into its own when microprocessors and memories appeared on the scene. Where do the insights come from that let people build the truss-and-trestle bridge, let alone the suspension bridge? Apparently, one of the answers is from recognizing what properties the new material has. For example, steel will take tension as well as compression, so it allows you to do new and different things.

Microelectronics differs from older electronic technologies, first of all, in being universal. In the olden days, you would have computers built with magnetic core memories, transistor or worse logic, special devices such as pulse transformers to shape pulses, and so on. Today, the nucleus of the system is all created out of one technology. There is no escape into any specialized technology for some special purpose.

Microelectronics is severely communication limited. Most of the area, power, and delay is caused by the wires. The transistors and active circuitry take up just a little bit of the lowest level. For reasons that I do not have time to get into, some of which are described in CSTB's *Computing the Future* (1992) report, these communication limitations favor simplicity and concurrency. You can even present a mathematical argument that explains why. Another effect is that it is easier to reduce costs than to increase speed. You will recall that the circuit density improves quadratically in scaling, whereas speed improves only linearly. These are some of the reasons people are resorting to unusual measures to get more speed, particularly parallelism.

There have been an enormous number of success stories. If John Hennessy (another former member) had been here, he could have told you about RISC (reduced instruction set computing) and cache memories. This success story is another case of trying to fit technology to people's needs through engineering insights that respected the limitations and exploited the specific capabilities of silicon.

DISCUSSION

MICHAEL DERTOUZOS: I have a question for both of you. I often hear this question about cost, and I would like to have it answered by the experts. What if you really went after the cheapest possible reduced capability of microprocessor chips that would still make a computer possible—not a computer that can support Microsoft Word 6 and the programs of today, but something that is perhaps scaled back 10 or 12 years. If you go for minimum cost, is the changed material—the bottom line material—in the cost? Would we see it reflected in the chip, and then in the computer?

CHARLES SEITZ: I have thought about that question and, to be realistic, the problem is economics. My answer comes partly out of reading one of your books. There have to be good profit margins in this business because it is extraordinarily expensive to keep developing new products on such a rapid cycle. Most computer companies are having to redesign their products about every 18 months. You sometimes wish that the reckless pace of our field would slow down a bit so we could all stop and figure out what has gone on, but with this pace, the profit margins on items such as desktop machines have to be reasonably large. One of the things hurting the companies right now is that the margins on desktop machines are very small. So I do not think you can cut the prices significantly below where they are right now.

WILLIAM SPENCER: We do an analysis, and I am sorry I do not remember the numbers now, on product costs on silicon. If you look at personal computers or anything else today, a larger and larger percentage of the cost is silicon. A major cost right now—forget the profit for a minute, if you can break even—is for capitalization on equipment. The depreciation costs are growing so large that they are going to dominate things. As an example, next-generation exposure tools, the most advanced ones, are now being sold. You may buy one for $5 million up-front, nonrecurring engineering costs and $10.5 million for the machine itself. It will be good for maybe two generations of technologies, five to six years. In this country, it has to be depreciated over that five-year period. That is turning out to be the major cost, the fabrication facility itself.

ROBERT KAHN: There is a lot of good news and bad news in the stories that we have heard from both of you. On the good news side, things are going to get smaller. On the bad news side, it is getting harder, and even the design is going to get increasingly difficult quadratically or by the cube of whatever. There are clearly two ways that we can deal with this increasing complexity. One way is to make things smaller, so you get more into a volume. Or on a given feature size, make things bigger in an area or volume. The attendant challenges are large.

When I first got involved with VLSI design in the 1970s, we were talking about 4 or 5 microns at most in size. In 20 years, Bill Spencer's charts show we are now down to 0.35 micron, soon to be 0.25. This is a factor of almost 40 or 50, some numbers like that, that we have experienced over the past 20 years. Yet it looks as if the projections for the next 20 or 30 years are maybe another factor of 2 or 3 at very large costs, pending some real innovation that changes the linearity of these curves.

I remember Jim Meindl put a report together 10 or 15 years ago predicting that somewhere around the 0.3-micron range, transistors would no longer function, and that this was a natural block. Somehow we have gotten through this. I do not know where the current limitations are, whether it is at 0.1 micron or whether we will get down to atomic scale.

The question I have is this. In the future, what is going to have the same factors of 50 or 100 or 1,000 in scalability that can generate the real interest and excitement in this field? Are we just talking about a tenth of a percent here or a few halving of microns here and there?

SPENCER: From the technology side, I believe the cost reduction will continue for another quarter of a century. We do not need a breakthrough to do that. The physics of transistors looks as though they are good down to less than 0.05 micron. We can build transistors that switch at room temperature; we do not have to go to liquid nitrogen temperatures to do that.

It would be great if the people in the IBM Lab in Zurich actually had gotten room temperature superconductors rather than liquid hydrogen temperature superconductors because, as Chuck Seitz pointed out, the major problem in the future is going to be 10 to 15 to 20 layers of metal or interconnect that may not be metal. This is where the controlling features are going to be.

In the past, every time we have run up against a technology barrier, somebody at Bell Labs discovered reactive ion etching to get rid of wet chemical etching. Steppers came along and gave us a 10- or 15-year respite in mask fabrication. Ion implantation came along and got us away from diffusion furnaces, which were terrible on the statistical issue that Chuck raised. My guess is that there is an engineer or a scientist out there somewhere who has ideas about how we are going to break through these barriers, and this will continue to grow. I do not think we need a big breakthrough for 15 or 20 years.

SEITZ: If I may add a couple of things. We thought that the scaling limit was somewhat below a quarter of a micron and would be due to tunneling, statistical fluctuations in threshold voltages at reduced voltages, and a few other effects. Now we know the limit is somewhere below a tenth of a micron. Over the next decade, what matters are not effects such as tunneling, but more mundane interconnection issues. The fabrication processes are improved in all of the areas where there is the most leverage, such as adding more and more layers of interconnect, which is really what has been limiting for some time.

Besides Bob, if, in your view, we can get a mere factor of 10 or 100 in the next 15 years, maybe the software people can take up some of the slack.

MISCHA SCHWARTZ: I would like to focus on a severe problem that Bill raised when he first started speaking, and that is the question of the downsizing of basic research in physics and devices by the major corporations in the last few years. Those of us in the academic world have been very concerned about this. The question is, Who is going to build the hardware platforms 10 years from now that the software is going to ride on? You have mentioned that maybe universities can pick up the slack. We keep hearing this. This is the reason why we are looking at increasing the basic research activities, things that are 10 years or more out. If I look at trying to rebuild Bell Labs, it is like reassembling Humpty-Dumpty to me. I do not think we can put it back together again. Bob Lucky could give us his view on it. My 15 or 20 years at Bell Labs convinces me it is gone and it cannot be replaced. National labs, in my view, require a larger cultural change than the universities do. I think industry must undergo a major change. Sending someone to work at a university is now considered a positive step for your career, rather than a detriment. I think this is the best place for us to turn.

I think innovations will come from all sorts of places. We have got to be in a position to capitalize on them when they do occur. That is the second major problem we have had in this country. We are not very good at this. We do the first start-up. Then, when we have a success, we find that East Asian countries, or even Europeans, now manufacture things in high volume better than we do. I do not think there is a simple answer to this, but somebody said we badly need a policy in this country that says how we are going to address this issue, not only in the silicon area, but in all areas of research. I think we no longer have national labs like AT&T and IBM, TI, Phillips, and Fairchild. The national labs are gone, and we have got to find a way to replace them.

SHUKRI WAKID: If you believe embedded computing is going to be real or very distributed, then intelligent sensors are one way to go. This means you need designs for chips that are a lot simpler, much simpler and more applied. By the same token, if federal digital signal processing is going to take off—and people sometimes say it is a barrier to computing because it is very difficult to do—then it is going to force a need for simpler design versus a more complex design. Do you want to say anything about this reverse trend of complexity?

SEITZ: Many of the processing media, sensor outputs and so on, of military embedded systems use digital signal processing chips. The typical digital signal processing chip dispenses with all the address translation hardware and other facilities required to run an operating system on processors used in desktops. There are at least 40 companies in the United States that make digital signal processors, ranging from boutique companies to the likes of TI and Motorola. As long as there is some money to be made in this area, I think it will continue to be healthy.

5

Computing and Communications Unchained: The Virtual World

Leonard Kleinrock
John Major

Leonard Kleinrock

I am tempted to suggest to the information technology users of the world, "Unite and throw off your chains." So let's do that.

I am going to talk about "Computing and Communications Unchained: The Virtual World." We could have called it the "virtual universe," but we thought we would restrict our focus. After all, academics are narrow (wasn't that the word used this morning?). I will talk about the virtual world of the nomad, discussing some of the issues and the technology. John Major will then talk about the applications in which you begin to see nomadicity happen.

So let us start at the right place, a dungeon. Most of us associate our computers, such as they are, with some kind of desktop device, possibly connected to a server, located down in somebody's dungeon. You never see it. You are rigidly attached to that architecture.

In fact, most of us are nomads. I do not know how many laptops there are in this room, but there are more in your hotels or automobiles. We travel everywhere. We travel to our office, home, airplane, hotel, automobile, branch office, and bedroom. My wife will not let me use a laptop in bed any more. What bothers her is the noise of the keyboard: it is forbidden.

Of course, we also travel to places like this symposium. When I go on the road, I usually travel with a laptop computer, pager, cellular telephone, and personal digital assistant (PDA). It used to be that shelf space in a bookstore or counter space in the supermarket was a precious commodity that all merchandisers would fight for. Today, they are fighting for waistline space to hang all this stuff on. You can rent out square inches of your belt these days. When I load up and go out into the world, I feel like Pancho Villa going into battle with the U.S. cavalry.

So what is nomadicity? In my mind, it is basically the system support needed to give all kinds of capability to nomads—no matter where they go—in an integrated, transparent, and convenient fashion. You should not have to suffer because you are moving around. Today, you suffer a great deal, so let us talk about what is needed and what some of the issues are.

Why should we care about nomadicity? There are many reasons. It is here right now; the users see it. We all move around and experience problems with synchronization, updating, access, weight, and more. I think there is a major paradigm shift in computing. It may not be as important as information warfare, but I think this is a dominant trend, not just a tangential issue. This is how people are now and how they will be using information technology. The technology is available. Wireless is here as well as the necessary light devices. Batteries are improving. Laptops are becoming more functional, if not lighter, and so on.

Nomadicity is multidisciplinary. There are no Renaissance people around who can deal with all the technologies involved here. You have to start off with nanotechnology and move all the way up to multimedia applications, including everything in between, to deal with this world. A multi-institutional effort or, at least, a multidepartmental effort, is required. Almost anything you do will be good because everything right now is bad.

You thought interoperability was a problem before. Now, however, people move around and suddenly appear 5,000 miles from where they last appeared on the net. Now what do you do with them? Who are they? Are they who they say they are?

Middleware is an area in which a lot of improvement is needed. Many people think of nomadic computing as wireless support alone, but this is just one component. You have all the problems and the fascination of nomadic computing without wireless ever entering the picture. Most vendors are putting out products that work in little niche markets and do not interoperate. I think that former Computer Science and Telecommunications Board (CSTB) member David Farber can attest to this. He always has a new device. These things do not interoperate. What is the mean lifetime of a device that you own before you replace it with something else, David? Two months. That is a David Farber unit, a DFU.

Since a lot of new research problems have emerged, anything you do will be of immediate, practical use. There are many reasons for getting excited about this field and understanding what is going on. In essence, the nomadic environment—your computing, communications, and storage functionality—should automatically adjust to everything you do in a way that is basically transparent to you. What components should be transparent? It should track you, for example, wherever you are. What you do should not depend on where you are located. You should always see a similar image.

What communications device do you have? Is it a PCMCIA card, a modem, or an asynchronous transfer mode (ATM) connection? Again, you should not have to think too hard at the user level. The system should adjust itself to your circumstances. How much communications bandwidth do you have? This is one of the biggest variables in the nomadic world. It goes from zero up to gigabits, and this can happen very quickly. It is not smooth in time or in number. The system has to understand this. You might see a difference in performance, but not necessarily a difference in what you have to do with the keyboard, or in your head, or with wires, plugs, configuration, or rebooting.

The ability of the user to be disconnected is one of the main attributes of nomadicity. You have to understand you are going to be disconnected, but you should not have to do much about it. You should act as if you are still connected. This would be the ideal. It requires that the necessary systems support be provided.

Also, you should be able to perform operations now, behave as if they are happening now, but have them actually happen later. For example, if I update a file, it may not actually happen for 10 minutes or 4 hours, but I have updated it as far as my operations are concerned. It should be independent of the particular computing platform that I have and, of course, whether or not I am moving.

This is where wireless comes into the picture. If I want to use computing and communications while I am moving, I need something like wireless. All the other issues come into play and have an impact without the need for wireless. Suppose I am sitting at my desktop computer, and I make a simple move and walk to my conference table and sit down at a laptop computer. I have now made a fundamental nomadic move. My platform is different. My communications are different. My screen is different. The keyboard might be different. Yet the environment that I would like to see should not change that dramatically. I should not have to say, "Oh, now I am in this operating system, and I have a different thing to do here." It should travel with me. Notice it is a 10-foot move from my desk to my conference table. How many elements on Charles Seitz's world map would that be? How many transistors cross over in 10 feet? Less than one, and yet it is a nomadic move. I would like that to operate not only in an office, but worldwide. I should be able to move around and have all the functionality follow me.

If you change your view, today's systems treat radically changing connectivity and latency as a mistake. This is a failure of the system, an exception. We design our systems to handle it as an exception. In a nomadic environment, this is the usual case, and you have to design your systems from the beginning to handle it. You have to assume you are going to be disconnected or facing radically changing latency often.

What are the components of the system design you have to worry about? Well, here is a list of the usual components: bandwidth, latency, reliability, error rate, delay, storage, interoperability, user interface, and cost. No surprises here; these are the usual suspects that we always worry about. However, there are some other things you have to worry about when you move to a nomadic environment: physical size, weight and processing power (how many Pentiums can I put on my lap?), battery life, communications, interference in the radio world, and damage. Moreover, these technologies are portable, which makes them easy to lose and subject to theft. I am sure you have read about the latest European scam that has now hit the United States. You are at an airport. You pull your luggage with the laptop attached. You reach the x-ray unit at the security gate. One person goes through the x-ray machine and his cohort is directly in front of you. When the cohort gets there, he fumbles and takes a few minutes to get through. Meanwhile, your laptop is on the way through. The first person gets it and disappears, and you have lost your laptop. This is a very easy way to lose a machine. Do not take your laptop to airports unless you watch where it goes.

Nomadicity exacerbates several concerns: disconnectedness; variable connectivity, either because the world does it for you or because you choose to move and use some other communications medium; latency, which varies significantly; variable routes in virtual circuits as you move around; variable requirements that you put on the system—what you expect and need; replication of resources such as files, machines, databases, and applications because you are moving; and foreign languages.

When you go to a new environment, you have to locate the power supply. Where is the modem? Where is the high-resolution screen or printer? You must become familiar with the environment. Conversely, the environment has to become aware of you. It must know that you are there. It should know your profile and what you want and should send the things to you that you expect. As the bandwidth and the platform capabilities change, the system should adapt what it sends to you. For example, it should compress some things and not send you high-resolution video. Perhaps it will send only the name of the movie, if all you have is low bandwidth. This is a major issue. Most of all, nomadicity is one of the ultimate problems in distributed systems.

What should we do? First, develop the systems architecture and network protocols. We need to interoperate between the wireless and the wired worlds to handle the concerns of unpredictable user, network, and computing behavior and to provide graceful degradation in all of this (simple problems!).

Second, we should develop a nomadicity reference model. We have heard today about one of the reference models that CSTB produced (Figure 5.1).[1] It has various names, the hourglass-shaped open data network (ODN), with a control level referred to earlier today as the bearer service. The ODN could provide a reference model. Suppose I wanted to send some e-mail. In one environment, I might connect to a cellular bandwidth source. A few minutes later, I might find that I can use a modem. Maybe I will walk into an office and get a 10-megabyte local area network, and maybe I will get a 150-megabyte ATM (if that ever appears in the office). As I move among these choices, what do I have to do? Well, right now I have to do a lot of things. I have to plug in a different PCMCIA card, reconfigure, reboot, and put down some IP addresses. I do not ever want to have to do this. I want something to do it automatically for me. This is one of the things that a nomadic system design should be able to take care of.

Another architectural model might be a more standard one—network infrastructure, support, and middleware. Hopefully, we will achieve integrated nomadic support as opposed to velcro integration, which we have today in many of our belt-laden devices. We happen to have one of these models at the University of California at Los Angeles. We filled in some of these architectural pieces (Figure 5.2). It has happened in a number of locations where you provide connectivity management, some file synchronization, and update schemes.

What else needs to be done? You have to understand how the system works and develop performance models. There are a lot of choices here. You can make a mathematical model, but the mathematics is not that strong. You can do numerical evaluation, but then you run into an exponential explosion of computation. There are iterative solutions, but then does the darn thing converge? Simulation—it is hard to search a large space. Emulation—it is

---

[1] Computer Science and Telecommunications Board, National Research Council. 1994. *Realizing the Information Future: The Internet and Beyond.* National Academy Press, Washington, D.C., Chapter 2.

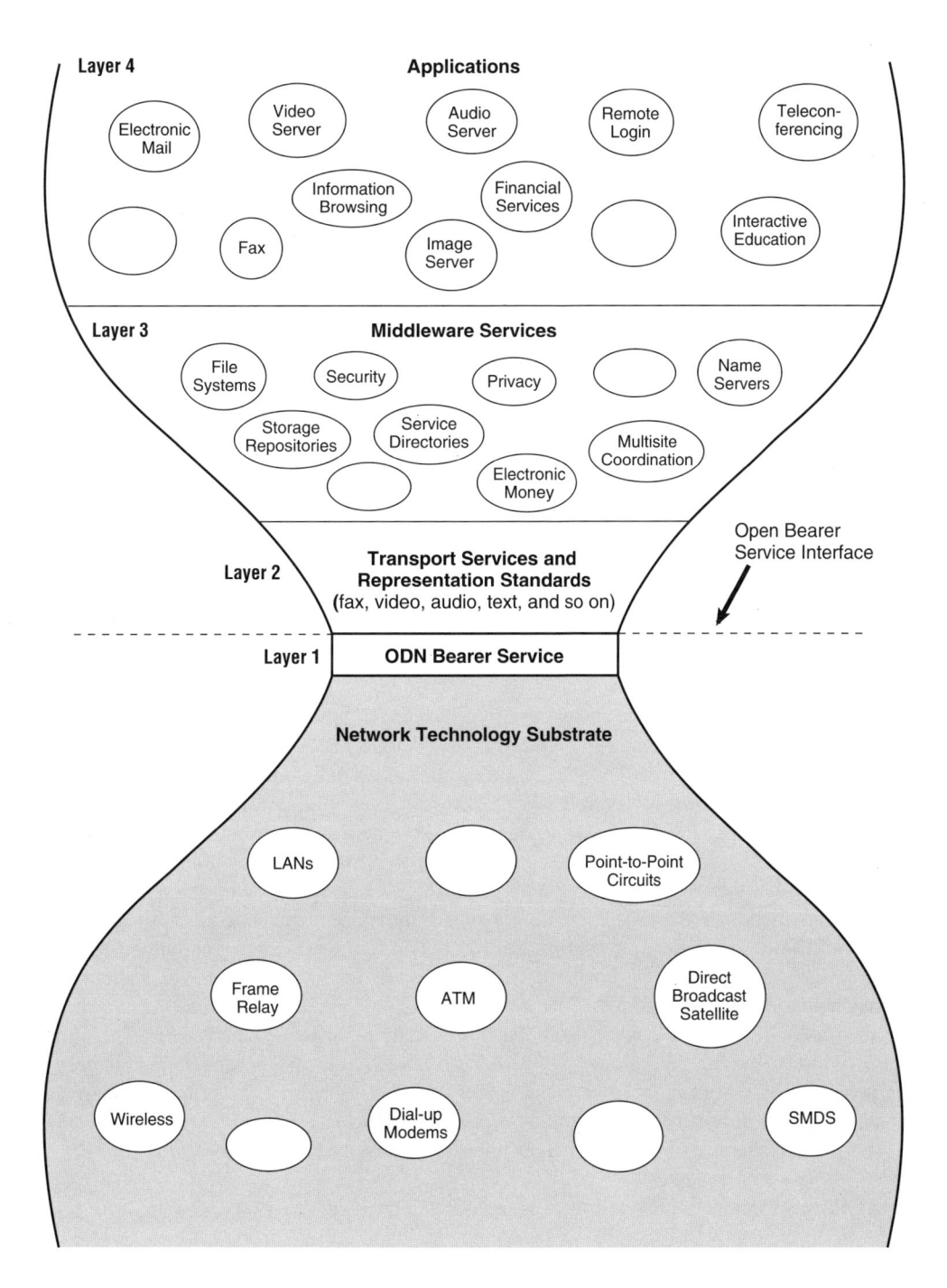

FIGURE 5.1 A four-layer model for the Open Data Network. Reprinted from Computer Science and Telecommunications Board, National Research Council. 1994. *Realizing the Information Future: The Internet and Beyond.* National Academy Press, Washington, D.C., Figure 2.1.

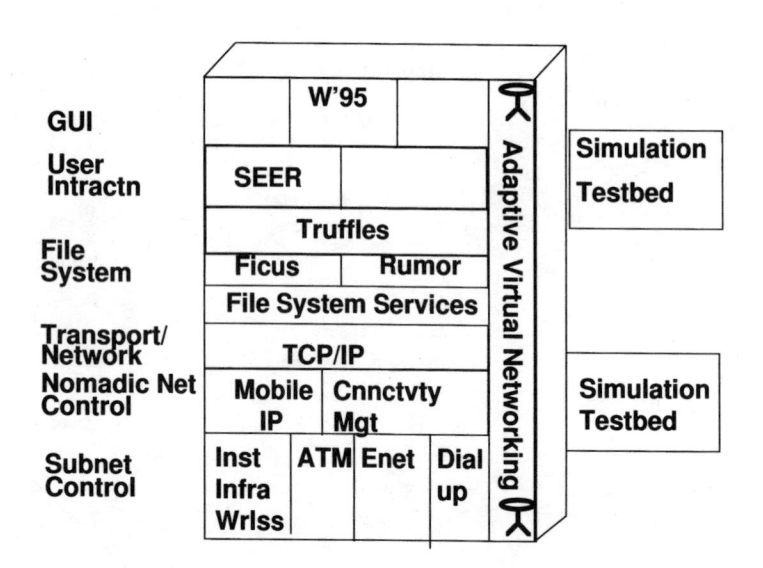

FIGURE 5.2 Architectural model for integrated nomadic support developed at the University of California at Los Angeles. Reprinted with permission from Leonard Kleinrock. Copyright 1996 by Leonard Kleinrock.

sloppy, ugly, and expensive (building the system and measuring how it behaves are certain to bankrupt you). The right answer is probably some hybrid mix to do the part that works best in each environment.

Consider adaptive agents (Figure 5.3). The classic assumption is that there is a big fat network between client and server. However, it is not client-server, it is client-network-server. The assumption is that you put a lot of stuff here because a big network will deliver whatever you need. In the nomadic environment, though, the network may be thin or zero. As the network skinnies down, you may want to move some functionality around in anticipation of a need for tools and data that you do not have with you. Adaptive agents at the middleware level are also called surrogates, proxies, helpers, and knowbots. We need a kind of theory or formalism, an architecture, a language for agents. They should help the nomads, the applications, the network, the servers, the communications devices, and the computing devices. They can sit everywhere. In a peer-to-peer application, or maybe a client-server application, agents may help to decide what goes on. We certainly need some of these adaptive agents inside the network as well to do the thinking for us, the compression, the connectivity management, and so on. This is the challenge.

So what do you observe from all of this? Nomadicity is here, you cannot escape it, and the needs are real. It makes every problem you ever thought about much harder. Each one gets an order of magnitude more difficult in this environment. It is a fascinating area (I think). The payoff can be huge. There is a severe lack of any integration, and there is chaos and confusion right now. You simply cannot afford to ignore the challenge of nomadicity.

Where will the next innovation come from? Ask your children—they will tell you, we won't.

John Major

I had a professor years ago who taught me that it is key to watch the moments in life when big things—new things—happen, because important things will follow. Words matter. When was the first time you heard the word *picosecond* referring to the speed of operation and *gigabyte* referring to storage capacity, or other words like those?

While Leonard Kleinrock and I were getting acquainted to do this presentation, he pointed out that he had just obtained a new laptop with two 1.2-gigabyte hard drives in order to have enough information when he is

- **Classical assumption:**

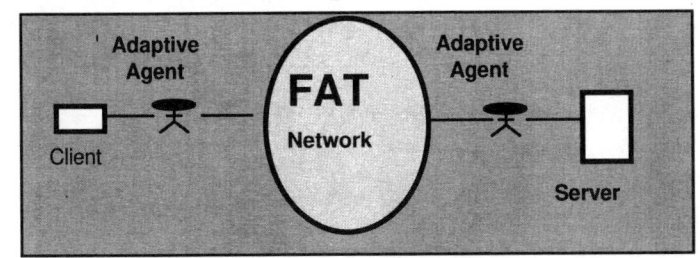

- **Nomadic Assumption:**

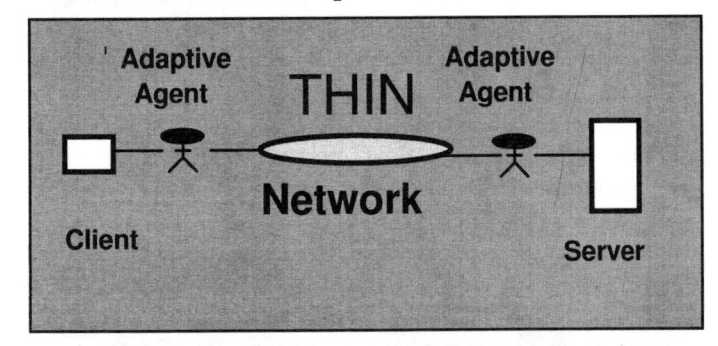

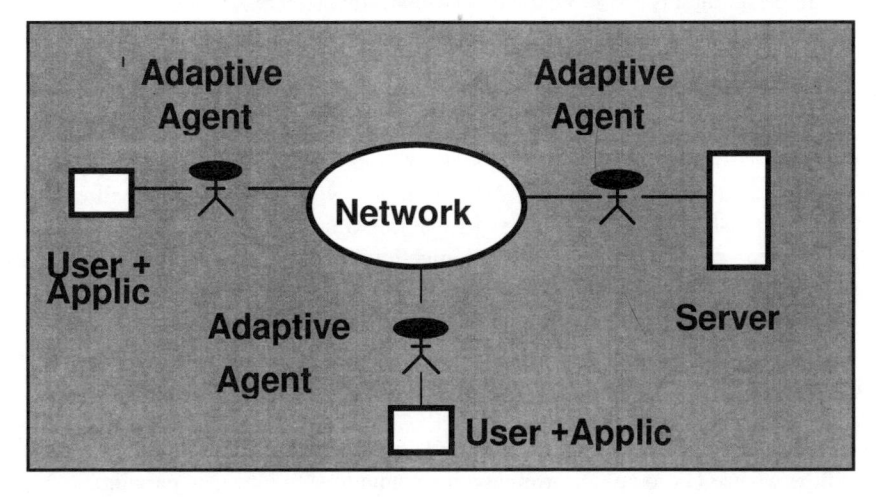

FIGURE 5.3 Adaptive agents. Reprinted with permission from Leonard Kleinrock. Copyright 1996 by Leonard Kleinrock.

"nomading" around. This was the first I had heard of two 1.2-gigabyte hard drives in a laptop, which is serious mobile computing. So I started to think that the vision I have of the virtual world will be well received.

I told Leonard about my concept of teleconferencing—that people at some point will travel virtually and no longer physically. In fact, that is the title of this talk, "Creating the Virtual World." With all sincerity he said, "That will never work!" So if I could not convince someone with two 1.2-gigabyte drives in his laptop that we are going to have a virtual world, perhaps I do not have much chance with the rest of you. The point is that this is a difficult concept. There is a lot to gain, but also a lot needs to be developed to make it possible.

Does anyone remember the 1987 movie *Planes, Trains, and Automobiles,* with John Candy and Steve Martin?

I travel a lot. I find it impossible to watch this movie all the way through because it hurts too much. Everything bad that has happened to me on one trip or another—not having money, the rental car not working, and the airplane schedules being unreliable—my whole life is being played out before my eyes in 120 minutes. It is a great movie, but it encapsulates every bad travel experience, and as such, it humorously captures part of the need for a virtual world.

You would think we would be highly motivated to change this. In fact, when I committed to this concept of describing the virtual world, I figured I would just go out, grab a few books, and catch up on the emerging theories. Then I would explain it to the audience, leave with loud applause, and feel very comfortable with myself. I discovered that, today, there is no agenda to create a virtual world. In fact, on some level, most of us believe that what we go through when we travel is important, part of what we do, and necessary to our society. In other words, we feel that there is no other way.

Yet, we know that travel is inefficient and that it significantly lowers productivity. It also costs a lot of money. The last time I looked, the federal government was spending about $40 billion a year to maintain the highway system. Office buildings cost money and isolate workers from their customers. Meanwhile, we are competing in a global world. If I believe it is important to be able to walk in to see my Motorola team and rally their spirits and convey information to them, and I do this for the team in Schaumburg, what about the team in Bangalore? What have I done about our lab in Australia?

If I think that being physically there matters, I should never go home—simply fly forever. Well, you cannot fly forever. With a truly global company, you cannot be in enough places. All of your time would be consumed in airplanes. So there is something wrong if you really want to run a global corporation or have a global team be as productive as it should, unless you are able to create, in some way, this virtual world.

How do you do it? How do you eliminate physical travel and commuting? How do you enable virtual travel and commuting? I will offer that it is not going to be via e-mail, fax, teleconferencing, or videoconferencing as we know them today. These represent a beginning; they are important steps, but they will not solve the problem. They may not even be reflective of the form of the eventual solution. To illustrate this point, imagine if low-cost robots had been invented by 1940 and were ubiquitous. Then someone said, "I am going to invent the dishwasher." Can you imagine what that dishwasher would look like? It would look like a robot doing the human tasks for washing dishes. As it was, there were no low-cost robots, so someone had to be very innovative. Today, as a result, dishes are washed in the dishwasher very efficiently, very conveniently, and completely differently from the way people wash dishes. This is the kind of fresh thinking that may well be required for virtual travel.

We have already seen some examples in communications of how new technology and uses change old paradigms. Today we have e-mail, which is in no way an analogue of what we saw before in the old teletype and telegram era. It is a completely new embodiment of messaging—transitional, to be sure, but very different. Remember centralized fax? One of my favorite memories is the time I suggested that I should have a fax in my office so I could communicate better with my colleagues around the world. I got a stern note from corporate, pointing out that fax was something that we had to be very careful about. There was an office you could go to if you wanted to send a fax. It was a couple of hundred yards from where I was. This centralized fax analogue has nothing to do with how we use fax today. So we have been able to step into new paradigms, but they are always very difficult to envision. In 1976, Jack Nilles said, "Telecommuting is the substitution of telecommunications for travel." This was the first time that the term was used. In 20 years, not much has changed, but there has been some progress.

There is some work being done in this area. There is some hope. There is some evolution. BellSouth has made a prediction that by the year 2000, 25 percent of corporate employees will, in fact, be working outside their traditional offices. By then, we will have evolved to the point at which a substantial amount of the work product will be separated from the physical office.

There are signs of progress. The growing use of e-mail is one. We are, in fact, learning to work as a global team with the primitive tools we have today. At Motorola, we have 140,000 employees and 120,000 computers. Since I manage this activity, it begs another question I am frequently asked by the chief executive officer, How can Motorola have almost one computer per person when well over half of our people work on assembly lines and have nothing to do with computers at all? Part of the answer is that we use these computers to run our factories, but the majority of the answer is that we are committed to staying in touch on a global basis, so many people have more than one computer, and there is a server-intensive network to support such usage.

At Motorola, we will soon be able to send a message to anyone anywhere in the world for less than a penny. It is a new kind of connectivity and convenience. We also have more than 100 videoconferencing centers. Everyone hates them; they are not natural in use or delivered image, but they are used regularly and they allow an early form of global group connectivity.

There are people such as Leonard Kleinrock coming forward with concepts of the kinds of structures and operating systems needed in order to have this sort of nomadic use of information technology. This is a nontrivial and very key next step.

There have been other relevant inventions. People can now do three-dimensional imaging on PCs with a relatively high degree of clarity and in a relatively simple format. People are developing fish-eye cameras and the mathematical processing that supports them. With these, images can be gathered by fixed imaging equipment and then reconstructed to your view, not the camera's view. The technology also exists to do similar things with images being reconstructed to give a rolling view from a set of still images. There is a camera product available that, based on the way you are looking through it, allows you to determine how the camera focuses and frames its image. There is probably a signal here of something very important for the emerging virtual world. People are beginning to see the kind of image manipulation that will be important to a virtual world.

We are starting to see more teleworking. The United Kingdom claims that about 7 percent of its workers do some type of teleworking. This figure includes around 2 million Britons who earn their living that way. Governments around the world are encouraging this trend. In the United States, the Environmental Protection Agency has a goal to reduce automobile commuting by 20 to 25 percent. The 1990 amendments to the Clean Air Act required that any company in an urban area with more than 100 employees find a solution to commuting to reduce the commuting load.

Slowly, but surely, we are starting to see processes that will enable virtual travel and commuting to occur. BellSouth is involved in this in a big way. Many of us were in Atlanta for the Olympics. Atlanta has a big problem. Even without the Olympics, with its road system, you cannot get from one place to another today. If you add several hundred thousand people, it does not get better. So BellSouth took on the challenge of the Olympics and used it to develop a mind-set in Atlanta that there must be another way to get your work done. They ran huge ads showing Atlanta traffic as it is normally, and at the bottom they asked the question, How will your employees get to work during the Olympic Games, when they cannot even get to work now? These ads were designed to motivate people to care about how they were going to get their work done when they could not get to work. The solution is teleworking, and this vast experiment will advance the cause.

AT&T has also jumped on the telework bandwagon and begun to do some interesting things. It is pushing the concept: forget the traditional work environment— let us move to a nontraditional work environment. Its phrase is "anytime, anywhere." AT&T's idea is that a substantial portion of the work force no longer works in the traditional office, and this number is growing on the order of 20 percent annually.

I have an interesting example of this. The man who manages the sale of IBM computers to Motorola sells on the order of $50 million worth of equipment each year. Clearly, this is a very successful account executive. Still, he does not have an office or a secretary. IBM has embraced teleworking to make its sales teams more effective. He works out of his home with his laptop. If he does not spend time in his office, where does he spend his time? He spends his time in our offices. He is talking to his customers and making himself more effective. This paradigm could not have existed in 1990, but it exists today and is successful. IBM is pushing the flexibility, lack of boundaries, and portability that will make this model very successful.

What are the drivers for change? First, the technology is making it possible. I had the pleasure of being at the Geneva Telecom Conference when Intel's president and chief executive officer, Andrew Grove, presented the process of global telecommuting and global networking—children talking to children around the world (today's version of pen pals in a global village), doctors in rural hospitals talking to doctors in urban hospitals. It is very powerful and an early indication of what we might be able to do.

Corporate restructuring and reengineering are also important factors. At Motorola, we do not have the luxury of providing a secretary for each individual. We can no longer afford to move everyone who is on a team into one location. We find, in some cases, that it is more efficient to have people stay at their normal workplaces. We have some early data related to the work of our software teams. In Fagen method software reviews, we analyze the software for potential errors before we actually write it. We want to find the mistakes before we create them. This is a fairly standard, emerging process in software; it is done as a team. We found that if we have these meetings on camera, they are more effective (by on camera, we mean that people can even be in adjacent cubicles; the screen is now split so that each of these people can be seen, as opposed to being in a conference room). That is, people work better out of their own workplaces, meeting in a virtual meeting room environment. Admittedly, what satisfies the needs of the software engineer for interpersonal contact and what would satisfy the same needs of people in other lines of work might be very different. This is, however, at least a signal that there may be something of value here.

At Motorola, we have shifts in labor pool availability. One of our key software teams works in India. If we flew the team back and forth to the United States for weekly or monthly meetings with other teams, there would be very little time left to get any work done. Clearly, we need a better answer, and teleworking is providing the answer with e-mail and high-bandwidth videoconferencing. Additionally, there is the problem of how to retain skilled resources. Sometimes the person you want to transfer or hire does not want to live where your business is located.

There are other examples of progress. Today, AT&T claims to have 40,000 part-time telecommuters, 12,000 mobile workers, and 5,000 full-time at-home workers. I, myself, am trying to stay out of the office one day a week.

Now, how do we get the job done? How do we create the virtual world? First, someone, everyone, has to start believing. People have to start getting a sense that all this travel is insane. People need to say things such as: "I am sitting in my fifth traffic jam this week." "I have just passed 200,000 frequent traveler miles; United loves me, but my family does not remember me!" The waste of time and resources and the pollution created can and should be stopped. Someone has to start saying, "This is wrong." There must be a sense that there is good reason to change. It will not happen overnight, but this consensus is building.

Then we have to start allocating to virtual highways a portion of the $40 billion that is currently going into physical highways. If 10 percent ($4 billion) were taken off the top and spent at universities, it would be a start. With this resource, the universities could spend time figuring out how to bring together these anecdotal developments to effect change. In addition, we must continue to encourage "green" legislation to discourage physical commuting.

The federal government, particularly the Army and Air Force, is running massive global organizations. It would be useful to plant the seed in people's minds that these organizations would run more efficiently if they had a new technology. Similarly, global companies such as Motorola must continue to challenge what the technology can deliver, as well as begin to develop and market products that can deliver the virtual environments that are needed. It is time to start convincing people, companies, and government that they would be more efficient if they had better tools.

These are my thoughts on creating the virtual world.

DISCUSSION

NEAL LAURANCE: I think in some sense you are trying to swim upstream. In the course of the management development in the past few years, we have seen management practice go from a hierarchical "find out what the boss wants, and then everybody get in line and do it," toward a much more collaborative environment, where you try to make use of everybody's talents and get everybody moving in the right direction. This process of getting everybody moving in the right direction takes a lot of human interaction. This has been the real limitation in trying to use some of the technology. Maybe what we need to do is to focus on what substitutes for body language in this virtual world. How can we convey this and get the consensus building process really captured in here. What are your thoughts on this?

JOHN MAJOR: I absolutely agree with you. We cannot step back. We are not going to give up teaming. The point I tried to emphasize is that teams are no longer located within 50 feet of your office. My team is located all around the world. Motorola is in no sense atypical in this. Many of our companies have massive investments in countries around the world. So I do not think we should shrink from the magnitude of the problem. We must understand that interaction may be a critical piece of it. There is surprisingly little work at understanding what the issues are.

ROBERT BONOMETTI: A few weeks ago, the Federal Communications Commission had what I think is going to become a historical notice of proposed rulemaking whereby it proposes to take 350 megahertz of spectrum and allocate it for a so-called national information infrastructure band or a supernet band or a hyperband, depending on whose buzzword you like. What are your thoughts as far as how this may be a critically important enabler for nomadic types of computing?

LEONARD KLEINROCK: Well, the problem of getting at the nomad when he is out of harm's way—for example, near fiber or wire—is a major one. Certainly the wireless spectrum provides a solution, be it satellite or local wireless. I think this is very important. I am anxious to see how the technology buys into it and what the devices are. This kind of freeing up of spectrum is important, but I do not think it is the most important problem. I think the systems issues are far more difficult and have to be developed first, but this is a component.

DAVID FARBER: A couple of months ago the dean at the University of Pennsylvania described me as his first virtual professor. He said it with pride in front of the board of trustees. With respect to John's comment about there needing to be some people supporting change, I think you are slowly beginning to see this when a largely liberal arts institution can absorb the concept without collapsing.

Let me make a couple of comments and then ask a question. When I travel (I am probably one of the most gadget-prone people around, courtesy of John and others), I tend to have a lot of gadgets. Half of my luggage consists of power supplies, batteries, everything incompatible with each other. So I would argue that there is a real need for standardization. It is pretty ridiculous when I have to carry five power sources so I can keep running.

I think that you alluded to a serious problem, Lenny, but let me see if I can draw you out a little bit. One of the big problems is the change of context every time I move from device to device, every time I move from one bandwidth package to another. I am sitting in an office, allegedly in the future, watching you via video link. I walk outside to my car and get on the road; I am on the way home, and I arrive at home. Each time that I pass through different bandwidth domains, everything seems to want to change. I think it is the discontinuity that I, at least, find very, very difficult to handle. I have to carry too many models in my head. Do you think this is going to change?

KLEINROCK: I think it will improve. It will never be as ideal as you would like. There will be ghostly images of me following you one way or another, be it by voice as you are driving or via a still picture of my unshaven face as you are shaving. There will be components like that. I think you are quite right. The discontinuities in bandwidth should be smoothed out somehow. One can anticipate this. One can, in fact, cache some of it. Predictive caching is a big component here, and it is very hard to do. This is why I say that there are a lot of good research problems, but these issues are key.

MAJOR: I think Len's concept of the intelligent agent that rounds off some of the edges and, in fact, modifies the communications environment automatically in response (where you simply get the headline and then can probe for the rest of the article) is the kind of concept that could be really contributory. To me, it is a fresh thought that the middleware community has not touched so far.

PATRICE LYONS: I was interested in your reflection on trying to liberate a few billion dollars from local and perhaps national highway programs to support the development of infrastructure. Well, if you had a bag of money and were a local government official, what minimal elements would be at the top of your priority list to encourage the kind of things you are thinking about?

What comes to mind from some of my legal debates in the last couple of years is the question of what is the proper role for government in a hierarchy of certifying authorities? For example, if you are a lawyer working in a suburb, your main office is downtown, and you are discussing things of an extremely confidential nature, you would want to make sure that there was some minimal integrity in the communications pathway. Perhaps this would encourage more people to do so. What would be the top one or two things that you would think about?

MAJOR: I will suggest that I do not think there is a quick answer here. I think the first pile of money has to go into basic research to understand the nature of the problem and free up the solution set from what would appear to be the direct path to what I described—to what might be a successful path to solving the real problem.

JERRY MECHLING: One of the problems we faced with the London project was about 200 to 300 people around the world who were not paid by us. They were like a network of different governments, universities, different entities—a complete mix. We were going to standardize on some collaborative work. We were going to do Lotus Notes. We decided not to, because the training requirements became too difficult and you had to fly them all someplace to get trained anyway. If you used video, you had to have all this language capability to translate, which was a problem because you had to explain the translation of the words. It became such a quagmire to figure out whether there was a way to standardize that we ended up with a network that was a complete hodgepodge of everything from PCs to fax machines to the most advanced systems.

Is there anything in the works that is going to address what I think is a problem people are not looking at: how do you conduct the collaboration when the players are not in the same organization? If it involves just one company, you would make a corporate decision, but you cannot do it in an interagency deal.

KLEINROCK: From a technology point of view, I cannot give you any answers to that question. I mentioned foreign language, the one you selected as a problem. Sure, translators will help, but they are a long way off. I am not sure what technology will overcome that problem. I think it is too early to standardize on a lot of this. We are just probing now; I would be frightened to do it at this point.

6

Picture This:
The Changing World of Graphics

Henry Fuchs
Donald P. Greenberg
Andries van Dam

Henry Fuchs

In this session, we will give you three snapshots of our favorite things in graphics—the views of three random folks that are not necessarily representative of the field. I will talk about displays, virtual reality (VR), and things that interest me; Andries van Dam will talk about user interfaces; and Donald Greenberg will talk about rendering.

The first thing I want to talk about is graphics in a nutshell since the 1980s (Box 6.1). Here is my mantra: 2D (two-dimensional graphics) has become ubiquitous. That is, 20 years ago, 2D was the cult; 10 years ago it hit the mainstream; and now you cannot get a machine without 2D graphics on it. If you want to get a 24 by 80 text-only screen, you have to try really hard. In some sense, what has happened to 3D (three-dimensional graphics) is 10 years behind 2D. It was still a cult 10 years ago; now, of course, you do not need to be in the cult of 3D in order to do 3D graphics in your work.

Some people say that VR, immersive displays, and interactive and immersive environments are also on this road, but I am a skeptic.

Speaking as a person who has been doing VR for 25 years, I think that the best recent development for VR is the Web coming along to take VR off the front pages. Until about a year ago, we had to spend a lot of time—and I suspect a lot of you who are working anywhere near the field spent a lot of time—just shooing people out of the lab in order to get anything done. In a nutshell, VR has been around doing useful work—by useful, I mean that people are willing to pay money for it when they do not have any alternatives—since at least 1970, in terms of flight simulators.

In the past 10 years, the military has spent lots of money on training both individual pilots and large groups of people, since it is not practical to train for all situations with real equipment. Also, the entertainment area is always going to be around, but with a nebulous future. However, I want to give you a couple of examples from medicine, an application area in which you can show incremental improvements that give some idea of what might be more widely available in the future (Box 6.2). Two examples can be presented using videos.

The first video is from Kikinis, Jolesz, and Lorensen.[1] Many of you are familiar with their work at Brigham and Women's Hospital in Boston. The interesting part of this work is that they are using graphics, and the merging of graphics with the patient, as a standard procedure now—not just in the planning of neurosurgery, but as a guide

---
[1] Videos were shown during the symposium.

> **BOX 6.1 Computer Graphics Since 1980**
>
> A. 2D became ubiquitous
>   - Desktop metaphor
>   - Desktop publishing
>   - Computer-aided drafting
> B. 3D: From cult to mainstream
> C. Hot topic: Virtual reality—so widely claimed as to encompass nearly all of interactive 3D graphics

or a road map for the surgery itself. This is a planning system for neurosurgery that, of course, starts with computed tomography (CT) or magnetic resonance imaging (MRI) scans. The novel aspect is showing the results registered on top of the patient. This guides the surgeon in deciding where to cut the hole in the skull and, once inside, how much tissue to remove.

Now we have systems that combine graphics with something in the outside world. My theme today is that we could push that capability forward for the rest of us. The next video provides an example of how to do this for applications in which the medical imagery is coming to you in real time. For guidance, you could use not only the old data, but real-time data as well. This is similar to the previous application, except now you are using the real-time imagery as part of the display in order to guide what you are doing.

In the next 10 years, we will find that the use of 3D graphics is increasingly commonplace. There will be a merging of the live video with the display. You might well ask what we need live video for in our everyday lives. I hesitate to tell you, but I think it is going to be the year-2000 version of the picture phone. I say this despite the fact that some of you might laugh, because picture phones have come around about every 10 years and so far have been disappointing to just about everybody. I expect that people are not going to be disappointed 10 years from now. There will be enough bandwidth, display capability, and computing to make teleconferencing a compelling shared presence, which the current generation of teleconferencing hardware cannot do. My personal view is that the crucial aspect of shared presence is going to be the capability to extract a sense of 3D from one place and show it at another place in a way that makes participants feel they are together, even if they are thousands of miles apart.

We are going to have to do what you might call desktop VR for modeling and 3D teleconferencing. In 10 years, I think we will have the equivalent of, say, a dozen laptops packaged within a single workstation, with about a dozen times the power and display capability of a laptop (more pixels (perhaps 10 million), higher resolution, and a wider field of view). Think about a picture-window kind of display made up of, for example, 10 or 12 laptop-size screens and 10 or 12 cameras. These cameras will use the workstation's computational resources to extract a model of the small environment you are in—a little office cubicle, for instance—and transmit that model to another place. In this way, together with head-tracked stereo, you could share the sense of presence with the person you are talking to or the two or three people you might be having a teleconference with. I believe the principal problem remaining is how to display these remote environments, because their data representation is probably not going to be polygonal. This is going to be the most exciting frontier: modeling from imagery and warping the images in such a way as to make the output image seem more realistic. A lot of people are working on this. The 1996 SIGGRAPH conference, for example, featured many papers on how to take imagery, warp it, and correct it from another perspective, making the output image more like a photograph than a picture of a set of polygons.

> **BOX 6.2 Virtual Reality in Medicine: Examples**
>
> - Neurosurgical planning and guide
> - Visualization of internal anatomy from preoperative patient imagery
> - Merged visualization of real-time internal anatomy and external views
> - Surgical training with simulators

These are my conclusions. We are going to have some automatic model extraction from video sequences, but not model extraction in the intelligent sense. The system will not know what the object is, but it will know where surfaces are in your little office cubicle. We are going to have desktop VR-like environments of what you might call a 3D window into another world. In addition, and totally separate from this, we will have laptops evolving into wearable computers the size of pagers; we will keep graphic displays with us all the time as we walk around. Much larger, higher-resolution images in the form of tiled displays will become commonplace. The capability will be far beyond the same 1,000- by 1,000-pixel displays we have had for the past 10 or 15 years.

Donald Greenberg

What has happened in the computer graphics industry in the first 25 years that I worked in it? The first system I worked on was a General Electric machine, trying to simulate the lunar landing vehicle docking with the Apollo spacecraft. It was as far as we could go in the 1960s: there was no lighting model and the colors were assigned to 3D geometries. This was the extent of the complexity that you could display in real time to try to train the astronauts in a nongravitational environment. About five years later, the graphics pipeline started. Without going through the details, you have a model; you have to describe where you are looking at it; you do a perspective transformation; you convert that to raster operations; and in the process, you do the lighting, put it into storage, and then display it. Around 1972, researchers at the University of Utah developed the Fong model. They never could see a full picture; there was only enough memory to see 16 scan lines at one time. The model was polygonally based and able to show diffuse reflections, transparency, and specular highlight. That was the basis of the graphics pipeline then; it is the basis of the graphics pipeline today.

The next step occurred around 1979. Turner Whittick from Bell Labs introduced ray tracing, which bypassed the camera perspective in raster operations. That is, you would send the ray through every pixel in the environment, compute where it reflected, pick up the accumulated light contributions, put an image in storage, and display it. The time required for an image as complex as his first publicly presented image was measured in days on a VAX 780, a 1-MIPS machine.

We have progressed a lot; we can do pretty well today even with complex environments, generating a picture with, for example, 32 million polygons in it. We are trying to see what happens as we start to get into complex environments, because the old algorithms begin to break up. We can generate the 32-million polygon image I showed only on machines that have 768 megabytes of local storage.

Interest in making pictures for the movie industry led to advances over ray tracing. Using thermodynamics, we tried a radiosity approach. Instead of doing millions and millions of rays, we took an environment, broke it up into thousands of polygons, and solved the interaction equations among all of those polygons. This is not a simple problem because, in contrast to finite element analysis, where there are nearest-neighbor problems, this approach relies on the global interaction among every part of the environment. There is a fully populated, nondiagonally dominated matrix that must be solved. In complex environments (for example, an occupied meeting room), the solution can exceed hundreds of thousands of polygons. Once solved, if it is for diffuse environments we no longer have to compute the lighting and we can use the graphics pipeline.

So where are we headed? Well, part of the problem we have had—and it is our own fault—is the fact that we have been too successful in making good-looking images. We cannot really discern whether the picture we are looking at is real or synthetic. If we think about what is happening now and what will happen in the future, we have to look at graphics rendering, at least, as a modeling of the physical world. Once we have modeled the physical world, we have to create the image of the physical world, and then we have to see how the human mind would evaluate it in the perceptual world.

Beginning with a model of the physical world, which has not only the geometry but the material properties on a wavelength basis (the light on a goniometric basis with its full distributions), we can do the correct energy distribution throughout the environment and find out what radiance is coming from every surface in the environment. Then we want to create the picture from where the camera is. Our camera really is our eye, which is made up of curved lenses, retinas, and so on. We have to create the perceptual image, taking into account the perceptual response and the physiological change. Then there is the interpretation, where the cognitive is not necessarily just what you see in the image; this is a tough problem.

So how do we start? Well, we take off our computer science hats and go back to Maxwell's equations from the 1800s. We try to get a simulation of the reflection model, which includes the angle of incidence, the wavelength, and the roughness and distribution of the surface properties—how much it absorbs and how much it reflects. We get an arbitrary distribution in directional diffuse and specular, a very complex five-dimensional problem. We have to do this: we want to have the model because we do not even have the measurements of most materials.

I want to go back to a comment that Edward Feigenbaum made. The problem we have in computer graphics is perhaps a problem in computer science in general; that is, we forget the engineering and experimental part of it. What we really have to do is test these paradigms in terms of their computation complexity. We have to do the experimentation to make sure that our algorithms are correct. We then want to distribute it, but in fact, we can distribute it right now only for diffuse surfaces because the rest is too complex when using the radiosity algorithms.

At Cornell, we have set up a $500,000 measurement lab in my laboratory to try to measure bi-directional reflectance distribution functions. We have come up with an approach using particle tracing, not the correlate techniques described by physicists years ago. Now we will send out billions and billions of photons, determine every surface they hit at every wavelength from every direction, and then try to accumulate this statistically (we now have bounds on the density distribution functions and their variance). Then we will be able to start to display these nonexistent environments by sending out these billions of photons. If we go into our measurement lab, build these environments, compare them, and then find they are correct, we will know that the simulations are accurate.

We also model the human eye—the reflection of the cornea, the lens, and the scattering of light inside the material so we can get the glare effects. We model the adaptive perceptions so that we see what something might look like in moonlight versus daylight when, in fact, our spatial and color acuities are very different, depending on whether the rods or the cones are involved. Then we build models, simulate them, and assess what is real and what is false.

So what is going to happen in the future? It is clear that as the environments get more complex, we are going to have more polygons. We are going to have in the neighborhood of 1 million to 2 million pixels, so the ratio of polygons to pixels or pixels to polygons is going to approach one. It started off with the pipeline. There were 400 pixels per polygon. Then it went to 100, to 50, and now down to 10. In the future, the balance point is going to be pixels to polygons, and I contend the graphics pipeline is going to go away.

Because we will have to disambiguate complex environments, we need all the perceptual cues, or shadows, of the interreflections. We will need global illumination algorithms, which are going to be common. As Andy van Dam said, we are going to need the progressive algorithms so we can get everything in real time. We will need to have a transition from WYSIWYG to WYSIWII. The former refers to "what you see is what you get." The movie industry is very happy with this because as long as the image is believable, the industry does not care whether it is real or not. I claim that it has to be "what you see is what it is" (WYSIWII) so that we can start to use these things to simulate what is going to be.

I would like to make three closing comments. First, with increasing processing and bandwidth, the real-time realistic displays—the "holy grail"—will occur within the next decade. If bandwidth is cheaper than processing and memory, then the client-server or cluster-of-computing paradigm will hold. If processing and memory are cheaper than bandwidth, then perhaps we will have local computation and display. In either case, however, we are going to have real-time, realistic image generation.

Second, I hope we will reach the stage soon where simulations can be accepted as valid tests. We do auto crashes that way. We do not build chips by building one and testing to see if it works. We test it on a simulator many, many times, so that on its first computation, it will work. At least in the design of automobile bodies or architectural interiors, for example, why can't we make sure that what we are going to see is what it is going to be? Again, the problem is that we have been too convincing in our images. If we start to believe in these tests, we could also start to use our simulated environments to improve the inverse rendering approaches and some of our computer vision problems.

Third, I would like to talk about the good news-bad news problem. The good-news problem is that 90 percent of the information goes through our visual system. The bad-news problem is that seeing is believing, and once we see the picture, we are liable to believe it. Photography will no longer be admissible in court. Seeing will not be believing, and it is going to be dangerous when we start to mix the real and virtual worlds. A picture is worth 1,024 words.

Andries van Dam

I am going to talk about what I call post-WIMP user interfaces, WIMP meaning windows, icons, menus, and pointing. Why user interfaces? Because we finally discovered that having raw functionality is not sufficient. If you cannot manifest that raw functionality in a usable form to ordinary users, it is not nearly good enough. Industries have discovered this and now have usability labs. Unfortunately, academia is still lagging a bit. The user interface does not get the attention that it deserves in computer science curricula, but this is a separate matter.

I will talk about what lies beyond the WIMP interfaces that we are familiar with today. This was, of course, presaged by Michael Dertouzos's plea that we should be much more interested in usability than we are and that computers should be much more user-friendly. This, to me, is an ideal. It was, by the way, considered an idea of the lunatic fringe 20 years ago that graphical user interfaces (GUIs) could make it possible for preschoolers—children as young as 2 years of age, I have been told, who can neither read nor write—to use computers productively. This is revolutionary, but it is just the beginning.

What else should we be thinking about? What is good about WIMP GUIs and what are they missing? What is good about them is ease of learning, ease of transferability from application to application, and ease of implementability, thanks to many layers of software. However, not every user fits that paradigm. Many people no longer want to point and click—those with repetitive stress injuries, for example, for whom speech is going to be far more important. Also, the layers of software support are both a feature and a bug. There is a huge learning curve in becoming a user interface programmer.

I am more concerned about the fundamental built-in limitations of WIMP user interfaces. They rest on a finite-state automata (FSA) model where the application sits in a state, waits for an input, and then reacts to it. It is half-duplex at its worst. Human conversation, on the other hand, is marked by the fact that it is full-duplex and multichanneled (that is, you see me gesturing while you are listening to me talking), and user interfaces do not mimic this at all. Key presses are discrete, but if we are involved in gesture recognition for handwriting or speech recognition, all of a sudden input is continuous and has to be disambiguated. The problem becomes far more difficult as soon as we move outside the boundaries of the user interface.

The WIMP GUI does not appeal to our other senses. It just uses the visual channel for output and the tactile channel for input. This is not nearly good enough. We would not be able to communicate well if all we had was vision, so we need to appeal to our other senses. Henry Fuchs talked a little about immersive virtual reality. You are not about to carry your keyboard and your mouse into an immersive environment where you can walk around freely—where the computer is continuously tracking your body, head, hands, and maybe even your gaze, and is communicating with you in a style that is very different from the WIMP GUI you are accustomed to.

So what is a post-WIMP user interface? Clearly, it has to solve the problems of multiple parallel channels and multiple participants. Computing as a solitary vice is largely going to go away and be replaced by teamwork and collaboration supported by computer tools. A post-WIMP user interface will require much higher bandwidth than we are used to for keyboards and mice. It will be based on probabilistic, continuous input that has to be disambiguated (consider today's experiences with handwriting for personal digital assistants or data gloves for virtual reality); it will need to have backtracking. The objects, unlike our desktop objects that just sit there and have no autonomous behavior, will have their own built-in behaviors, and we will interact with them as they are interacting with each other. Twenty years ago, the Media Lab at the Massachusetts Institute of Technology had a wonderful demo ("Put That There") of somebody who could talk, gesture, and interact with objects on the screen that did their own thing; this is the kind of future we are moving toward.

WIMP GUIs are not going to go away; they will be augmented, suitably and appropriately. I see a spectrum of user interface techniques. On the one hand, there will be direct control—I, as a user, will be empowered to tell the application what I want it to do. On the other hand, there will also be indirect control where I can dispatch

> **BOX 6.3 Challenges for 3D User Interfaces, Virtual Reality, and AR**
>
> - Much better device technology is needed.
> - Design discipline is necessary.
> - A powerful development environment is required.
> - "Driving applications" are needed that make productive and cost-effective use of the technology and drive it.

viruses or agents, however you view them, that will operate on my behalf. They will communicate with me in reasonable ways to let me know what they are up to so I can adjust them and make sure that they are not harmful. Agents and wizards in primitive forms are already here; you will see them used far more in the future, for better and for worse, especially on the Internet. Finally, we are going to have speech and gesture input.

A different set of user interface techniques based on 3D graphics holds promise. These involve 3D widgets and a sketching user interface that employs gesture recognition rather than traditional WIMP techniques. The 3D widget is a specialized tool that lives in the same 3-space as the 3D application object. The sketching interface provides a more limited vocabulary of direct manipulation.

There are a number of challenges that still have to be met before any of these techniques are used routinely in industry (Box 6.3). One that I can talk about briefly is time-critical computing, which the networking community calls quality of service. Instead of having algorithms that compute perfectly and take however long they require, we want algorithms that yield a usable result within a given time limit and produce higher-quality results if given more time. This will enable us to schedule the number of frames that we are able to generate and avoid motion sickness. Such time-critical computing requires a new way of looking at algorithms, something the artificial intelligence folks have been doing for a while. Object models are another pet peeve of mine. All of the current object-oriented programming languages have static, class instance hierarchies. In the 3D graphics multimedia world, we need dynamic, evolving objects that can change their type and their membership on the fly at run time. This is not supported, nor of course is any kind of input or output, which is all relegated to a library.

In conclusion, things are going to change pretty dramatically in the future. Change will be driven not so much by what is happening in our university research laboratories as by the driving forces of the entertainment industry, which is putting enormous resources into making this revolution happen. In the graphics community, we are all very keen on these new technologies, but we need to conduct usability studies to find out whether they are really useful. This is terra incognita, but it is a very exciting new frontier.

DISCUSSION

WILLIAM PRESS: I have a question for Donald Greenberg. An idea whose time comes and goes and never really arrives is eye tracking, because people have estimated that the real bit rate into the entire visual cortex is perhaps only a megabyte. Does that have any future now?

DONALD GREENBERG: It definitely has a future. We are making a lot of progress on it. We are also looking at the difference in the full view versus peripheral vision, which means that we can reduce the amount of resolution we need on the slides. Henry is conducting the particular tracking experiment.

HENRY FUCHS: You characterize it accurately: it is something that comes and goes but has not really arrived. I think it is like the picture phone. An optimist will say that when all the necessary technological components have arrived, then it will have some advantages. I think it has not arrived because various pieces of the puzzle are not in place yet.

Even when we are able to do eye tracking, for example, we are unable to take advantage of it because of other limitations such as the display mechanism itself, the graphic systems, and various other things. My guess is that, in 10 years, there will be a few selected applications, but it will take at least 10 years until we see some part of it have an impact.

7

Computational Biology and the Cross-Disciplinary Challenges

Deborah A. Joseph
Edward H. Shortliffe

Federal Research and Funding Policies

Deborah A. Joseph

I will begin by talking a bit about the breadth I see in computational biology as a discipline, and then identify some of the challenges I see. Discussions at the Computer Science and Telecommunications Board (CSTB) have made me realize that some of my more computationally oriented colleagues have a narrower view of biology (and computational biology) than I have. Some did not take a biology class after high school, while others graduated from college before DNA sequencing was something that every undergraduate biology student learns about. Computer scientists are well aware that new technology has caused rapid advances in computer science and telecommunications. However, new technology, both biological and computational, is also rapidly influencing work in biology.

Computer scientists often equate computational biology with algorithmic support for genomics and molecular biology. Work in this area is fairly well known by computer scientists, and the computer science conferences often contain papers on genome sequencing and mapping. Computer scientists are fairly familiar with the problem of predicting the secondary and tertiary structure of biological molecules, and applications in drug design. The artificial intelligence community had done some nice work using neural nets and hidden Markov models for finding genetic indicators and genes within genome sequencing data. Problems such as predicting gene linkage are somewhat less well known, but are receiving increasing attention in the algorithms community.

However, beyond the areas of genomics and molecular biology, computational biology is much less understood by computer scientists. Asked what other topics of research are found within computational biology, most computer scientists would probably think back to its origins within mathematical biology. Problems such as population modeling and developmental models (how the zebra get its stripes) would come to mind. Some of the most exciting new research in biology that is benefiting from new technology and computational methods is almost unknown in computer science. I will mention just a few examples.

Currently, an exciting area of research is the quest for a better understanding of life in extreme environments. This work includes the study of deep sea vents, hot springs, extremely acidic environments such as mine tailings, the ice of the Arctic and Antarctic, as well as the possibilities of extraterrestrial life. Studying the organisms of these environments in large part involves studying their DNA. A particular gene, the 16S rRNA gene, has played a key role in their classification. Computational biology has been central in providing algorithmic techniques for constructing phylogenetic trees based on molecular sequence data for this gene. These phylogenetic trees can then be used to predict the evolutionary relationships between known and newly discovered organisms. Using these techniques scientists have recently discovered an entire new domain of Archaea (bacteria). Some of these organisms appear to be evolutionarily very old and may offer clues to understanding the origins of life.

In other research, computational biologists are building models in areas of biology, ecology, and animal behavior that have typically been very noncomputational. Computational models in some of the areas offer important new possibilities, as well as controversy. For example, new models of animal physiology may in part replace some toxicity and carcinogenicity studies previously done using animals. The acceptable role of computer simulations and models for biological risk assessment is an area of ongoing, and at times heated, discussion.

Computer modeling has been used for many years in agriculture to predict crop yields. More recently an entire area of information intensive agriculture—precision agriculture—has developed. This area brings together many tools: variable-rate applicators, global positioning systems, databases, plant growth models, and models of nitrogen and soil characteristics to improve the way we raise crops. Through total farm management systems we can hope to make farming more efficient and to reduce nonpoint source pollution. Better models of farms as ecological systems will help maintain them in a healthy, efficient, environmentally sensitive way.

The last area that I will mention, tools for research and collaboration, is more familiar to computer scientists. Many of the database and visualization tools developed outside of biology are now being imported into biological research tools. Consider the work I discussed earlier on the construction of phylogenetic trees from molecular sequence data. As one might suspect there are a variety of algorithms that can be used and they sometimes give conflicting results. Today there are visualization tools being developed that will allow one to morph one phylogenetic tree into another. This gives the researcher a way to dynamically, rather than simply statically, visualize the different trees presented by different algorithms.

Systems are also being developed to aid in experiment design and hypothesis testing. Tool kits for building biological models are helping to eliminate the need to go back and start programming from scratch each time. Lastly, large database systems for managing experiment design take the typical laboratory notebook and put it into electronic form. They make it possible to take the data from a large experiment (a genome sequencing project, for example) and make it a flow-through process.

The topics I have mentioned are broad, but are certainly not all-inclusive. Today computational biologists can be found working in all areas of the biological sciences. Just as broad as the problems pursued are the techniques employed. They range from numerical approximation, mathematical programming and numeric solutions of mathematical equations, to neural networks, discrete algorithms and data structures, databases, and visualization.

I have given a very brief overview of some of the exciting work being done today in computational biology. At this point I would like to discuss some of the challenges. Although my talk was to focus on funding challenges, I will broaden what I say to include some more general challenges that I think the field faces. As one might expect, these are frequently interrelated and are challenges that many interdisciplinary fields face.

Federal funding for research in computational biology comes from the National Institutes of Health, National Science Foundation, the Department of Energy, and in more narrowly focused programs from DARPA, the Environmental Protection Agency, the National Institute of Environmental Health Sciences, and the Department of Agriculture. Industrial funding is significant in some areas of biotechnology and agriculture. Within the federal agencies there are two models for funding. One is a model exemplified by the Computational Biology Activity at the National Science Foundation: a specific budget and review process to handle research awards in computational biology. The second model makes funds available for research in computational biology through disciplinary research programs in the biological sciences. Those programs that benefit from particular tools and algorithms are expected to support their development. So, for example, research on modeling of biological molecules might be handled through a program in molecular biology. Each of these models of funding have advantages and disadvantages. I will discuss some of these as I talk a bit about some of the general challenges that computational biology faces.

The first challenge that I will mention is that of disciplinary identity. Computational biology is often described as forming the bridge between biology and computer science. But, is there more to computational biology? Does it form its own discipline, have its own foundation, its own fundamental problems? As scientists, we often judge fields by their foundation, and by the difficulty and fundamental nature of their research problems. If as computational biologists we describe our research accomplishments solely in the disciplinary terms of computer science and biology, computational biology will fail to develop as an independent discipline. This will have an impact on funding in computational biology and make it difficult to build a cohort of scientists with the interdisciplinary training necessary to push the frontiers of research combining computation and biology.

A second problem is closely related. As scientists, we typically credit depth over breadth, and interpret this as brilliance over diligence. Scientists that work in computational biology must learn the background science and language of both biology and computer science. Further, scientists that work as part of an interdisciplinary research team pay a price in time as communication across disciplinary boundaries (and the physical boundaries of work places) is often slow. Scientists that work in a narrowly focused area simply have more time to delve deeply into their own science. We reward the time spent on science in a way that we tend not to reward the time that a scientist spends learning to speak a different scientific language and interact with other scientists. This affects funding and advancement in computational biology as it comes out in our review process.

A third funding challenge is what I will call the "glass ceiling" for computational biology. When there are lots of resources or there is a special program in computational biology, funding is not a problem. However, when computational biology lacks its own funding program or when projects grow and outgrow their special program, computational biology projects must compete in disciplinary programs in biology. Many will say that this is as it should be; the disciplinary programs in biology should support the computational research that they benefit from. Nevertheless, when funding becomes short these projects are often the first to go. Even in the best of times, it is difficult to find funding for large computational projects within disciplinary programs. I believe that as computational tools become a more routine part of biological research some of this problem will disappear. Nevertheless, cooperation between funding programs can also ease some of the strain.

The last issue that I will raise is what I will call it the "hacking problem." Whether we are mathematicians proving theorems or computer scientists implementing large software projects, much of our work in science is incremental. We build on our past experiences, our past knowledge and the past knowledge of others. Most new results extend this chain of knowledge; only occasionally are directions radically altered. How we judge this type of incremental research varies. When results in computer science are extended to produce new results in computer science, the review of this work may be quite different from that when a computer scientist applies a novel (but known) date structure in a new way within biology. Too often the cross-disciplinary work is viewed in the negative as "applications" work or "hacking." The ambiguity between implementation and novel extension of an idea often causes problems in the scientific review process.

Although these challenges exist, I am optimistic about the future for computational biology. There are exciting problems to be worked on and an excellent group of students interested in the area.

Finding a Home in Academia

Edward H. Shortliffe

As someone who works at the intersection of two fields, medicine and computer science I have thought a lot about bridging issues that affect interdisciplinary fields, of which computational biology is only one example. The more general discipline that deals with all biomedicine plus computing and communications, known as medical informatics, is the field in which I have worked personally over the years. In recent years, the role of bioinformatics has become increasingly important, and I will tell you a little bit about computational biology as it fits within a medical informatics training and research environment at my own institution.

The challenge of finding a home in academia is my nominal focus today. However, I am going to broaden my charge to discuss issues related to how people who are trained in these intersecting fields find professional outlets in the world beyond, whether in academia, industrial research and development, or other settings where leaders increasingly are recognizing the need for people who have insights and experience that cross disciplines. Much of what I am going to say today harkens back to the CSTB report *Academic Careers for Experimental Computer Scientists and Engineers*, which was completed in 1994 (Larry Snyder from the University of Washington chaired that committee); it addressed the issue of how experimental (as opposed to theoretical) computer scientists best fit in this evolving world of computer science within the academic community.

As the field of computer science broadened, we began to hear comments like one that I overheard at a faculty meeting in the computer science department at Stanford in about 1988, "I remember the days when I used to go to departmental seminars and I could understand every talk!" This remark was made by an eminent computer scientist, reminiscing about the good old days when the computer science department was first formed and he could go to any colloquium, understand the content, and ask insightful questions. He realized that it was getting harder to do that. He no longer bothered to go to all of them because some were in areas in which he knew he would have trouble relating to the topic. As computer science both broadened in scope and developed detailed methods within narrowly defined subareas of investigation and application, and as seminars focused on subtopics in areas of specialization, it is hardly surprising that some of these subspecialization areas would involve overlap with other disciplines in which computing was being applied. Biomedicine has been one of these.

One might argue that there is a potential scientific discipline at the intersection of computer science with almost every other field because of the ubiquitous nature of computing and communications as they touch on all aspects of society. Therefore, it is not surprising that we are seeing the evolution of expertise in many of these areas of overlap. The question is, To what extent does interdisciplinary training need to be specifically provided? To what extent can people get good training in computer science and simply choose to apply it in a given area? Or to what extent can people get well trained in the areas of application and then try to pick up enough computing so that they can work effectively at the intersection? Given the complexity of computer science and the tendency either to reinvent the wheel (if you have not trained in computer science) or to apply computing notions naively (if you do not really know the discipline of application), many of us believe that there is an ever-growing need to identify the disciplines that lie at the intersection and to train people explicitly in these areas.

As you might imagine, this was the first question I was asked when, in the early 1980s, I proposed the creation of a new interdisciplinary degree program at Stanford in medical informatics. Understandably, faculty in the computer science department said, "Well, you know we could have a lawyer come in here tomorrow and ask the same question, and then the nutritionists will come in and say that computers and nutrition is a new field, and they need a degree program as well. Where do you draw the line?" Such questions do force you to identify what is unique about an interdisciplinary field that is not being addressed adequately during training in one area or the other in order to justify creating an entirely new degree program. We have, accordingly, tried to characterize the specialized knowledge that lies at the intersection between biomedicine and computer science. Figure 7.1 shows one perspective of the challenge of trying to understand biomedical knowledge and biomedical data, to get them into computer programs, and then to use them for purposes of inference and problem solving. These processes define a set of research areas that explain the focus for much of the current training and research in medical informatics.

I am going to tell you a little about the Stanford informatics degree program to give you a feel for how an effort to train people explicitly at this intersection between biomedicine and computer science has played out over a 15-year period, where people are getting jobs, and some of the lessons that we have learned from this experience.

Our program offers master's and Ph.D. degrees, and it is based in the School of Medicine. The core faculty in the program all have both M.D.s and a Ph.D. in either computer science or medical informatics. We are trying to train people who will be researchers either in industry or in academic positions.[1] The program has close ties to

---

[1] Our program has grown to a steady state of approximately 26 students. About 60 candidates apply each year for four to six positions. We have about 8 core faculty overseeing the program, although there are some 35 faculty in the university who are involved in student projects or have volunteered to be preceptors for students who want to work in their areas.

Since 1984, we have received a National Library of Medicine (NLM) training grant that supports both pre- and postdoctoral degree candidates. The postdocs are essentially all physicians; they are considered postdocs by the university, but they are all graduate students, too. The resulting anomaly (postdoctoral trainees who are also degree candidates) confuses our university (including its administrative computers) in that we have a group of students who are postdocs in the eyes of the medical school and graduate students in the eyes of the university. Some of the university's policies about the two groups are conflicting. The predoctoral candidates tend to be either medical students who are getting joint degrees or students straight out of undergraduate school. The latter group does not necessarily want to get a medical degree or another health science degree but does want to make biomedical informatics its professional career commitment. Trainees who are not supported by the NLM training grant receive support from regular research grants, and some, especially international students, bring fellowships with them.

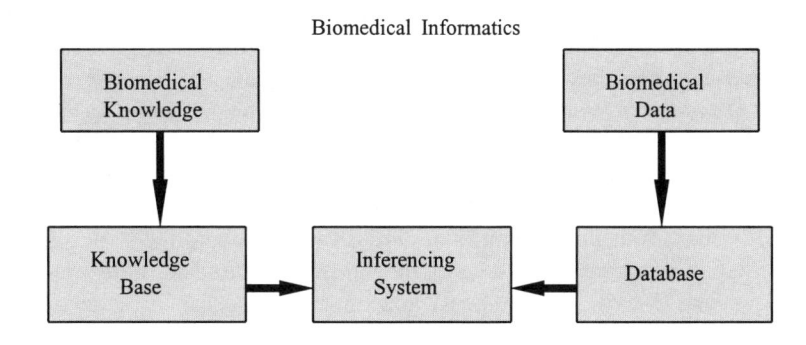

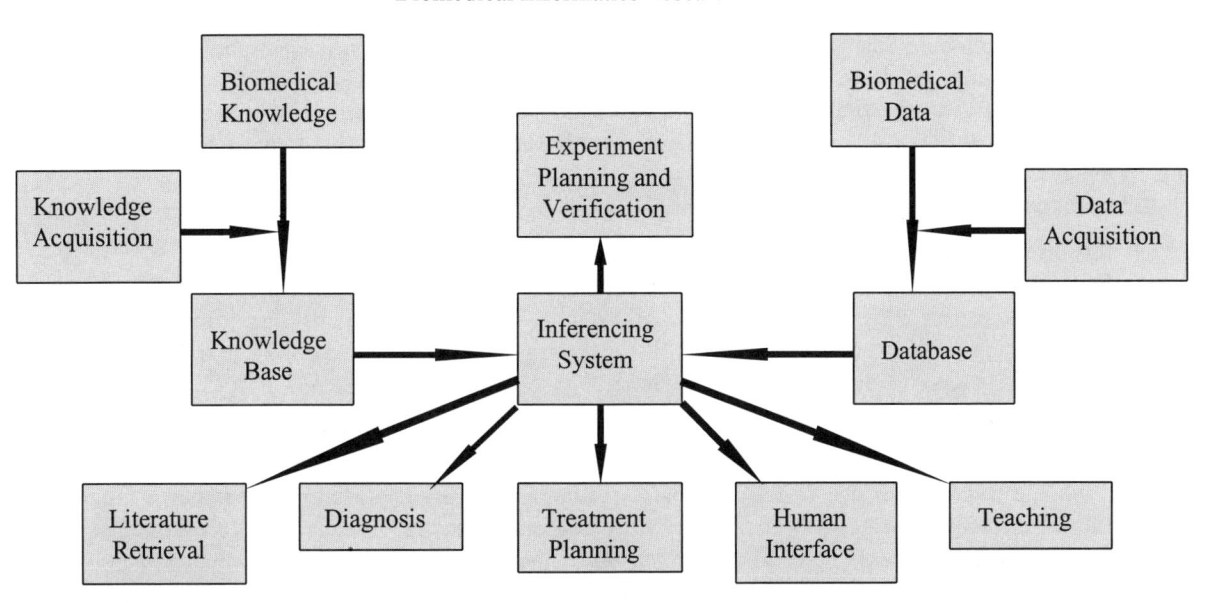

FIGURE 7.1 Elements in the process of incorporating biomedical date and knowledge into computer programs for use in problem solving. SOURCE: E.H. Shortliffe. 1996. Stanford University.

the computer science department; we happen to be on a campus where the computer science department and the medical school are close to one another. This greatly facilitates the opportunities for interchange between the two environments, allowing students to take a medical school course, hop on a bicycle, and take a computer science course in the next segment of the day's class schedule.

Our emphasis is on rigor coupled with methodological innovation. We decline to give degrees simply because a student builds a clever program, regardless of its size or complexity. Instead, we ask that our graduates be able to describe what the scientific issues are that generalize from the project they worked on to contribute broadly to the work of others. We try to make people think about what they are doing in this context so they can stand up before a computer science colloquium or national meeting and describe their work, playing down the biology issues or the medical issues and focusing instead on the computer science, and be perceived very much as peers when it comes to the nature of the contributions of their work.

Similarly, the same project often needs to be explained in a medical grand rounds, a pediatric seminar, or a molecular biology research seminar. When the same student goes into that setting, he or she needs to be able to play down the computer science, which the typical person in the audience will not understand, and focus instead on what is exciting and challenging about the biology being done in those projects. This is quite a set of demands to place on people. It is one of the reasons our students tend to be older and actually end up with dual training. About two-thirds of our students are physicians, who then take this kind of degree as their subspecialty training after getting a medical degree and often after completing a residency program.

Since there are many practical issues in biomedical computing, we need to find the right balance between teaching students the theory and giving them the skills that allow them, when they graduate, to have an impact in applied settings. The challenge in building a curriculum for this kind of interdisciplinary field is to think about all the areas students need to know something about (see Figure 7.2). As the bottom of Figure 7.2 indicates, we believe students need to learn something about both clinical medicine and the basic medical sciences. Thus, computational biologists in our program take courses on genetic structure and protein unfolding, and they learn about clinical medicine topics as well. You might ask how much students can learn and retain in any one of these areas if they are trying to learn something in all of them. We believe the answer is that graduate training requires a broad background across a set of fields that one is likely to need to know about, with subsequent focus on an area of specific application or methodology during the period of one's M.S. practicum or Ph.D. dissertation.

In addition to computer science and biomedicine, we add courses in decision sciences, biostatistics, bioengineering, and epidemiology. If people take courses in all of these areas, it is not the same as taking a computer science degree and then applying it in biomedicine. The curriculum provides an unusual mix that did not previously exist in any single degree program. As more schools offer degrees in this area, we are seeing a similar mix of topics being included in their curricula.

The program philosophy rests on the following:

- It is oriented toward research training for academic or industrial careers.
- It has close ties with, and gains inspiration from, the computer science department.
- There is emphasis on rigor, methodological innovation, and an ability to generalize from specific results.
- There is emphasis on verbal and written skills, including an ability to present work to three audiences.
- Training is balanced between theory and practice.

Accordingly, we have five major content areas in our curriculum: biomedical informatics, computer science, decision science, biomedicine, and health policy. Biomedical informatics courses include the incremental offerings that we had to create for our degree candidates. Among these are courses in computational biology, an introduction to medical applications of computing, medical decision support systems, biomedical imaging, and a project course in which students build small systems under the guidance of medical informatics faculty. Our informatics courses are cross-listed in the computer science department, and many computer science and engineer-

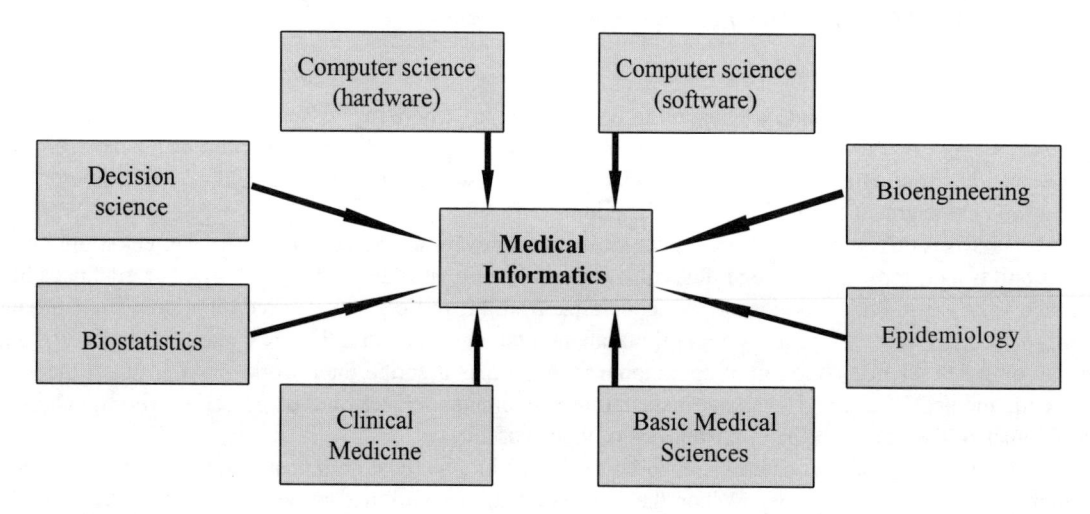

FIGURE 7.2 Areas of knowledge required in the field of biomedical informatics. SOURCE: E.H. Shortliffe. 1996. Stanford University.

ing graduate students are coming to the school of medicine to take our offerings. They see biomedicine as an interesting area of application that they often had not realized could be pursued as part of their academic program until they discovered our courses in the computer science portion of the university catalogue.

Let me summarize some of the lessons learned from this training activity. First, the trainees themselves often become tremendous bridge builders. They have brought informatics preceptors into contact with faculty throughout the university that we would not otherwise have known because the students take courses all over the campus. They come back with ideas, oftentimes with strong recommendations about people to whom we should be talking. Sometimes this has played out in very positive ways, resulting in new collaborations and relationships, often outside the medical school. As I mentioned, we have people from other parts of the university taking our courses, learning about the field, and getting excited about it. Some of them end up going into medical informatics or going off to medical school. Although that may not have been their original plan, they realize that there is an emerging career opportunity at the intersection of health care or biology and computer science.

The crucial role of an advanced computing environment is probably obvious to anyone in computer science, but this point is not always obvious to deans of medical schools. It is difficult to provide high-quality training in medical informatics unless you have a computing environment that is just as advanced as one expects to find in a typical engineering school department. Budgets in medical schools seldom take this into account. Thus, we face the challenge of trying to build an advanced computing environment in a school of medicine, which does not have a tradition of supporting such environments from operating funds. We are very dependent on industrial gifts of hardware and software as a result. There is also a tension between the academic and service notions that exist within a school of medicine. Some medical school colleagues assume we should be pulling wires through the walls for them and making sure that their PCs work. Similarly, our affiliated hospitals, realizing that we have talented students who know a lot about computing in the health care setting, often wish we would provide them with technical support for their applications projects.

Some service activities of these types would be healthy for our students; many will be involved with precisely such issues after they graduate. On the other hand, if an academic unit becomes a service organization within a medical environment, it is difficult to maintain the academic standards that are part of the motivation for having the training program in the first place. Thus, we try to find the right balance and to avoid formal service responsibilities to the school or hospitals, just as computer science departments do not run their university computing centers.

With regard to our current status, we have enrolled about 70 students since we started the program in the early 1980s; 33 were physicians when they started, and another 13 were medical students while they studied with us.[2] Much more than half of our graduates end up as M.D.-Ph.D.s or M.D.-masters; the majority of these earn Ph.D.s. Essentially all of our current students, 24 out of 26, are Ph.D. candidates.[3] It has become clear that there is declining interest in master's degrees in this field, at least among our own trainees. One of the reasons there are currently more master's graduates is that it takes only two years to train a master's graduate and four to six years to train a Ph.D., as in most Ph.D. programs.

People often ask what our students do when they graduate. They do research; they work in academia; they work in industry doing research; they work in operational jobs within medical institutions or, increasingly, in biotech companies that are hiring people who have these kinds of skills. Some of them do clinical, administrative, and educational management. We have one or two that actually run hospital computing groups. Some are working on digital libraries, providing expertise in information retrieval and information science. Seventeen of our graduates are in academia, sixteen in industrial settings, and three in clinical practice, but still doing some informatics as part of their life as practitioners; one is in a hospital working in a clinical computing environment; two work for the government (one as a military chief information officer and the other as a researcher for the National Aeronautics and Space Administration doing biomedical computing); and five are completing their residency programs.

---

[2] Of the 70 trainees to date, 17 have been women, which reflects about the same ratio as our applicant pool over the years.
[3] There are 24 graduates with master's degrees; 20 with Ph.D. degrees.

- "Wet bench" science
- Medical schools
- Supremacy of discovery

- Creation of artifacts
- Engineering schools
- Supremacy of methods and theories

Molecular biology

Computer science

Computational biology

Not "scientific"?   Too applied?

Clinical medicine

Computer science

Medical informatics

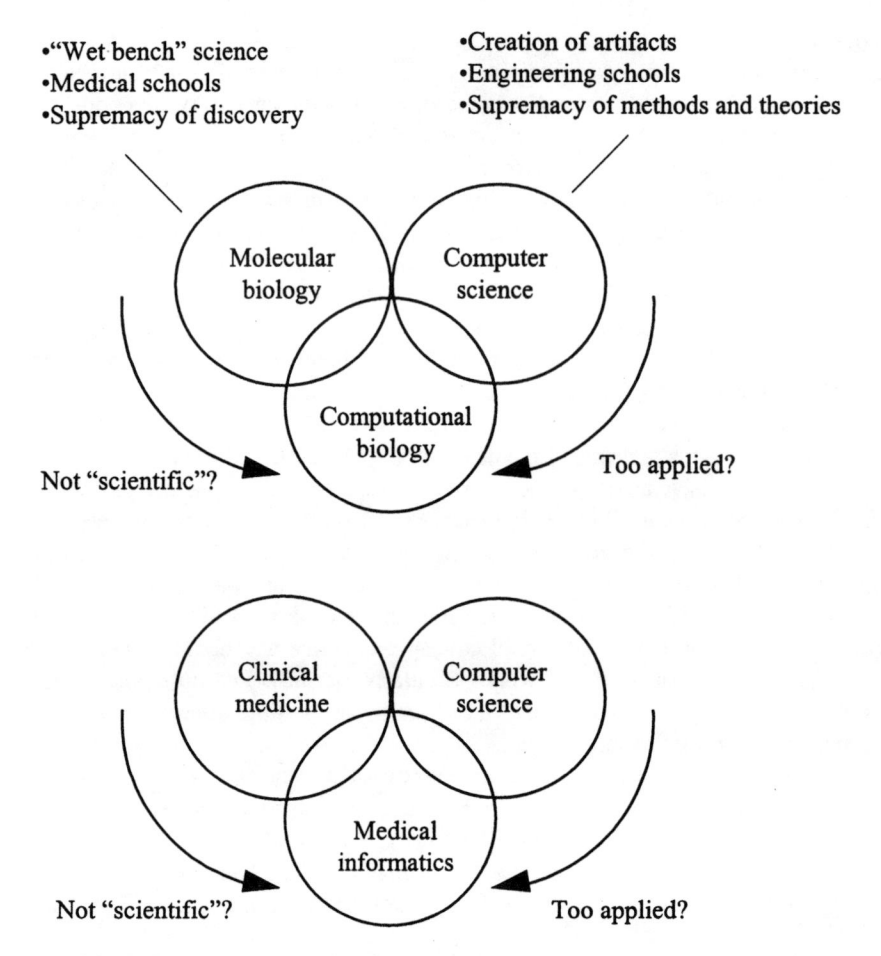

Not "scientific"?   Too applied?

FIGURE 7.3 Elements of interdisciplinary fields and the challenge of integrating them effectively. SOURCE: E.H. Shortliffe. 1996. Stanford University.

I would like to close by coming back to a point raised by Professor Joseph. Computational biology can be considered as providing the interface between molecular biology and computer science (Figure 7.3, top), and one can anticipate the emergence of experts who identify this as their professional field; yet such experts must try to maintain credibility in all three of the circled domains in Figure 7.3, moving as fluidly as they can among three rather different environments. The challenge is that oftentimes the molecular biologists ask whether what these people are doing is true science. Can you be doing true science if you are not working with test tubes and gels at the bench? In the culture of a biomedical research environment, it is common for colleagues to define "true scientific research" using such criteria.

Ironically, from a different perspective, computer scientists sometimes see computational biologists as being too applied. They are viewed as being driven too much by applications from the real world. They may not be working at the theoretical level that would more easily validate their role as traditional computer scientists in an academic setting. The same picture applies to those doing medical informatics (Figure 7.3, bottom). Although the relevant area of application is clinical medicine, medical practice and clinical faculty often view "valid" research as focusing on discovering explanations for life processes, with an emphasis on "wet bench" research. Such narrow definitions and high expectations have forced those who work in a field like medical informatics to define their discipline not in terms of their impact on applications but, rather, based on the significant research problems that are associated with the development of new methodologies and solutions. Perhaps all new fields, drawing on elements from existing disciplines while defining a new one, have had similar challenges in defining and explaining both the scientific content and the practical importance of the problems that they address.

DISCUSSION

DAVID MESSERSCHMITT: Let us assume that in some area such as computational nutrition we decide that it is not appropriate to have a separate program. I can imagine three models. We could have a nutritionist who learns enough computer science but is in the nutrition department, or somebody who is in computer science who gets in that application area, or someone with a joint appointment. Maybe Deborah in particular could comment on the relative merits of these three options.

DEBORAH JOSEPH: I guess there are two issues. One applies to a faculty member or researcher and the other to a student. One of the things that the University of Wisconsin has that I think has worked very well is a National Institutes of Health training grant in biotechnology. We do not offer a degree in biotechnology, but we nevertheless have students who are trained in biotechnology on this grant. These students have advisers in some aspect of the biological sciences and in some aspect of the mathematical computational or engineering sciences. The program has been extremely successful. Students tend to spend a lot of time together, and there tends to be a lot of cross-fertilization between students in the program. This would fulfill some of the need for a program in computational nutrition.

At the faculty and research level, I think that more of it has to do with how we view our colleagues, more than the barriers put in place by universities. The true impediments typically relate to whether you can get through some review process, whether your colleagues respect you, and whether you have colleagues who will write letters of recommendation for you. If we as scientists value cross-disciplinary activities, I do not think that good people who work in computational nutrition will have trouble. If their research is not valued by both their colleagues in nutrition and their colleagues in computer science, then they have trouble.

MESSERSCHMITT: Can they survive in either department?

JOSEPH: I think at the University of Wisconsin they can, but it depends a lot on the university.

EDWARD SHORTLIFFE: Local culture is very important in this regard. I think there is a great difference among institutions as to how much they are purists with regard to the implied criteria for appointment and promotion, as suggested in my last two diagrams. What we are talking about is credibility at promotion time, especially in the tenured line.

JOHN RIGANATI: One comment is related to the question you just had, but I wonder if either of you knows what the trends are in biology departments around the country about giving credibility to computational biology, as opposed to requiring that the computational aspects be linked directly with wet lab work. I know that a few years ago the percentage was very small, and I do not know if there have been any significant changes as a result of the kinds of things that you describe here. The second question is whether or not biological processes are viewed by either of you as having some basis for practical application in computational devices in the next few years, either literally or in an inspirational sense.

JOSEPH: I guess I could take the last question. Recently DNA and biomolecular computing have received a lot of press. I think we have been overeager in computer science to believe that this is a technology that is right around the corner. It is an interesting idea, but I will reserve judgment on whether we are really going to build supercomputers out of DNA molecules.

SHORTLIFFE: To answer the first question, there has been a growing trend in the recruitment of individuals who really know about computational biology to work in traditional molecular biology, genetics, and biology departments. Frankly, it is impossible now to do modern molecular biology research without local expertise in computing, especially for DNA homology searching and analysis of some of the protein databases and the like.

Molecular biologists can no longer do effective research unless they have somebody in their midst who can help them understand and use the newer technologies. Sometimes professors assign graduate students to go off, learn about computational biology, come back, and basically be the "guru" in the lab. Increasingly there has been a trend toward bringing in people who provide that expertise professionally in-house.

The problem is that they are not necessarily viewed as peers. I think the way we are beginning to see computational biologists perceived as pure scientists is in the creation of separate academic units within institutions. Here they are recruited as computational biologists, they are evaluated as computational biologists, they are promoted as computational biologists, and then they become collegial collaborators with people in other departments on research projects. The same goes for medical informatics.

MISCHA SCHWARTZ: The issues you raise are common to many interdisciplinary areas. There is one school of thought that says perhaps the way to do this is to learn one discipline well and become a specialist in it and then go on to the other one, rather than spreading yourself too thin. What are your comments on this?

SHORTLIFFE: Well, it was indeed the question I was asked most when we proposed our program. It depends on whether you really believe that there is a body of knowledge at the intersection that warrants full-time focus during training.

There is a difference between learning a lot about medicine and then learning a little about computing, or learning a lot about computing and then a little about medicine, and focusing your entire graduate training at the intersection itself. I believe there is also a significant difference between formal informatics training and having someone first finish a medical degree and then earn a computer science degree in a conventional computer science environment; the connections to medicine are not part and parcel of the way computer science is taught, and understanding the relationships and relevance is thus inherently left as an exercise to the student. We have found that there is a kind of culture in a field that students begin to absorb if they train in an environment where everybody else is also working at the intersection of disciplines that interest them.

I think time has shown that people who actually train explicitly to work at this intersection—individuals who get all their course work and culture in an environment that allows them to interact regularly with others who have the same interdisciplinary interest—develop a special skill set as bridge people and as productive contributors when they get out. Such skills are not easily acquired by earning one degree or the other and then sort of secondarily adding a pure version of what they missed.

WILLIAM WULF: I personally think that the issues raised here about interdisciplinary work between computer science and biology go across the board. Computing intersects absolutely everything, as Ted said. So this is a problem that we computer scientists really have to face up to. I think we can be a model for a lot of other disciplines where the same problems occur, but perhaps there is not quite the same urgency to collaborate at the moment.

8

Visions for the Future of the Fields

David D. Clark, Moderator
Edward A. Feigenbaum
Juris Hartmanis
Robert W. Lucky
Robert M. Metcalfe
Raj Reddy
Mary Shaw

Introductory Comments

David D. Clark

In contrast to the previous symposium segments in which people presented set pieces, the Visions panel was designed to be entirely interactive.

We talk about the future of the field, but it is "fields"—plural—because the Computer Science and Telecommunications Board (CSTB) deals with computer science, computer engineering (this assumes that you believe they are different things), and telecommunications, especially with the acquisition of the redefined "T" in CSTB's title. Some parts of electrical engineering are also relevant. In addition, we have many different kinds of players—academia, industry, and government—as well as user groups and spokespersons for societal issues. Our concerns are multidisciplinary even within individual fields.

We have a future that is shaped by a variety of forces. Within each of the relevant fields, the question is, What is going to happen? Is it going to converge? Is it going to fly apart? We could ask some technological questions about the future. Are processes going to get faster? Are networks going to reach the home? The more interesting questions, perhaps, are the societal ones that may transform the world in some way. These are actually the hard questions to answer. Another thing we can do, especially with regard to some of the societal issues, is try to bound the possibilities. What are the boundary conditions of the technologies?

Panel Discussion

DAVID CLARK: Earlier in the symposium, we heard the phrase "the reckless pace of innovation in the field." It is a great phrase. I have a feeling that our field has just left behind the debris of half-understood ideas in an attempt to plow into the future. One of the questions I wanted to ask the panel is, Do you think that we are going to grow up in the next 10 years? Are we going to mature? Are we going to slow down? Ten years from now, will we still say that we have been driven by the reckless pace of innovation? Or will we, in fact, have been able to breathe long enough to codify what we have actually understood so far?

RAJ REDDY: You make it sound as though we have some control over the future. We have absolutely no control over the pace of innovation. It will happen whether we like it or not. It is just a question of how fast we can run with it.

CLARK: I was not suggesting that we had any control over the pace of innovation, but are you saying you think it will continue to be just as fast and just as chaotic?

REDDY: And most of us will be left behind, actually.

ROBERT LUCKY: We were talking this morning about the purpose of academic research. The problem that many of us involved in research have is that, as at Bell Labs, we used to talk about research in terms of 10 years. Now you can hardly see two weeks ahead in our field. The question of what long-term research is all about remains unanswered when you cannot see what is out there to do research on.

Nicholas Negroponte was saying recently that, when he started the Media Lab at the Massachusetts Institute of Technology, his competition came from places like Bell Labs, Stanford University, and the University of California at Berkeley. Now he says his competition comes from 16-year-old kids. I see researchers working on good academic problems, and then two weeks later some young kids in a small company are out there doing it. You may ask, "Where do we fit into this anymore?" In some sense, particularly in this field, I think there must still be good academic fields where you can work on long-term problems in the future, but the future is coming at us so fast that I just sort of look in the rear-view mirror.

MARY SHAW: I think innovation will keep moving; at least I hope so, because if it were not moving this fast, we would all be really good IBM 650 programmers by now. I think what will keep it moving is the demand from outside. In the past few years, we have just begun to get over the hump where people who are not in the computing priesthood, and who have not invested many years in figuring out how to make computers do things, can actually make computers do things. As that becomes easier—it is not easy yet—more and more people will be demanding services tuned to their own needs. I believe that they will generate the demand that will keep the field growing.

JURIS HARTMANIS: As was stated this morning, I think we can project reasonably well what silicon technology can yield during the next 20 years; the growth in computing power will follow the established pattern. The fascinating question is, What is the next technology to accelerate this rate and to provide the growth during the next century? Is it quantum computing? Could it really add additional orders of magnitude? Is it molecular or DNA computing? Probably not. The key question is, What technologies, if any, will complement and/or replace the predictable silicon technology?

CLARK: I wonder if growth and demand are the same thing as innovation? Mary, you talked about a lot of demand from outside. We could turn into a transient decade of interdisciplinary something, but does that actually mean there is any innovation in our field?

SHAW: We have had some innovation, but it has not been our own doing. Things like spreadsheets and word processors, for example, that have started to open the door to people who are not highly trained computing professionals have come at the academic community from the outside, and they had very little credibility for a long time. I remember when nobody would listen to you if you wanted to talk about text editors in an academic setting. Most recently, there has been the upsurge of the World Wide Web. It is true that Mosaic was developed in a university, but not exactly in the computer science department. These are genuine innovations, not just nickel-and-dime things.

EDWARD FEIGENBAUM: First, I would like to say a few words about the future, and then I will pick up on the theme that Dave Clark started with, the debris, and ask some of my friends in the audience about their debris.

There has been a revolution going on that no one really recognizes as a revolution. This is the revolution of packaged software, which has created immense amounts of programming at our fingertips. We go to the store; we buy it. This is the single biggest change from, say, 1980. I think the future is best seen not in terms of changing hardware or increased numbers of MIPS (or GIPS), but rather in terms of the software revolution. We are now living in a software-first world. I think the revolution will be in software building that is now done painstakingly in a craftlike way by the major companies producing packaged software. They create a "suite"—a cooperating set of applications— that takes the coordinated effort of a large team.

What we need to do now in computer science and engineering is to invent a way for everyone to do this at his or her desktop; we need to enable people to "glue" packaged software together so that the packages work as integrated systems. This will be a very significant revolution.

I think the other revolution will be the one alluded to by Leonard Kleinrock, what he called *didactic agents* or *intelligent agents*. Here, the function of the agent is to allow you to express what it is you want to accomplish, providing the agent with enough knowledge about your environment and your context for it to reason exactly how to accomplish it.

Lastly, I will say something about the debris. I can bring my laptop into this room, take an electric cord out of the back (presuming I have the adapter that David Farber was talking about before), and plug it into the wall. I get electricity to power my computer anywhere—in Wichita, La Jolla, on any Air Force base in the country, or anywhere else I might be, even at the National Academy. Yet I cannot take the information wire that comes out of the back and plug it into the wall in Wichita or La Jolla or any Air Force base I choose because all of a sudden I need the transmission control protocol (TCP) switcher. I need to have exact contexts for the TCP to operate in those particular environments. We do not yet have anything like an information utility. Yes, I can dial the Internet on a modem, but this is a second-rate adaptation to an old world of switched analog telephones. It is not the dream. The architecture of the Internet—wonderful as it may seem—has frustrated the dream of the information utility.

ROBERT METCALFE: There are two solutions to your problem. The first relates to the structure of the Internet, and for this I must defer to Robert Kahn. Since he and Vint Cerf are the fathers of the Internet, they must answer this question. The second solution to the problem has to do with what Gordon Moore has recently called "Grove's law." Grove's law states that the bandwidth available doubles every 100 years. It is a description of the sad effects of the structure of the telecommunications industry, which would be in charge of putting these plugs where you want them. This industry has been underperforming for 40 or 50 years, and now we have to wake it up.

LUCKY: What was the question? We are pushing something we would like to call IP dialtone. I see this as the future of the infrastructure right now, to have an IP network. There was an interesting interview with Mary Modahl in last month's *Wired* magazine. They asked her if voice on the Internet would really take over, and she said no. She said that real-time voice is a hobby, like citizen's band radio, not a permanent application. I actually think that in the future, the voice may be a smaller network and the IP infrastructure will really take it over. IP dialtone will be the main thing. I would not rebuild the voice network. I would just leave it there and build this whole new network of IP dialtone networks.

CLARK: Part of what marks our field is this reckless pace of innovation into the future. Another is the persistence of stubborn, intractable problems that we have no idea of how to solve. An obvious problem that was raised earlier in various guises is (to look at it abstractly) our ability to understand complexity or (to look at it more concretely) our ability to write large software systems that work. When we go to CSTB's 20th anniversary celebration and look back, do you think we are going to see any new breakthrough? Let us pick this stubborn problem as an example; then we can talk about some others. In software engineering, is something actually going to change? Are we going to see a breakthrough? I am thinking about the point Ed Feigenbaum made that people are going to be able to engineer a software package at their desks. I said, "Oh no. It is done by gnomes inside Microsoft." Won't it be done by gnomes inside Microsoft for the next 10 years?

SHAW: I think this is a very big problem, and Ed pointed out a piece of it—that the parts do not fit together. We have, though, this myth that someday we are going to be able to put software systems together out of parts just like tinker toys. Well, it isn't like that. It is more like having a bathtub full of Tinker Toys, Erector Sets, Lego blocks, Lincoln Logs, and all of the other building kits you ever had as a child; reaching into it; grabbing three pieces at random; and expecting to build something useful.

As long as we have parts that are intended to interact with other parts in different ways and we cannot even recognize quite how any given part is expected to interact, we will have a problem. We do not even have distinctions explicit enough to do the analogue of type checking—to say, "This one does not fit with that one, what can I do about it?" Well, maybe there is nothing we can do, and maybe we can find a piece that will patch it up. I think this is one of the major impediments to being able to put together systems from parts and make them work. I do believe that we will be able to make progress. Breakthrough is a pretty big word, but I think we will at least be able to make significant progress on articulating these distinctions and helping each other understand when we have the problem and what, if anything, we can do about it.

The other problem that Ed Feigenbaum raised is the nonmigratory local context. I have the same problem that Ed does, except mine is at the software level. I put a document on a floppy disk and I take it someplace. Well, maybe the text formatter I find when I get there is the same one that the document was created with—how fortunate. Even so, the fonts on the machine are not the same, and the default fonts in the text formatter are not the same, and it probably takes me half an hour to restore the document to legibility just because the local context changed—I see everyone is nodding, so I can quit telling this story. Then, of course, there is the rest of the time, when I find a different document formatter entirely. This is another example of having parts that exist independently that we want to move around and put together. Once again, I think the big problem is not being able to articulate the assumptions the parts make about the context they need to have.

BUTLER LAMPSON: I say we just had a breakthrough. How many breakthroughs per decade are you entitled to? The breakthrough we just had is the Web. You had to cobble together a few million computers, a whole bunch of servers, all kinds of legacy databases and documents, and all kinds of stuff. All you have to do is write a few PERL scripts and you can patch together huge amounts of stuff and make it accessible to millions of people. What is all this whining and moaning about? Furthermore, I would like to point out that if you want your document to be portable, just write it in vanilla ASCII and you will not have any problems with portability.

SHAW: I am really good at ASCII, and ASCII art too, but we were planning the next decade's breakthrough.

CLARK: You used a portable operation object as an example. I actually think that is a lot easier than integrating software modules. I thought when Butler stood up he was perhaps going to say something about the viability of distributed object linking and embedding (OLE). Is this the answer to composable software?

LAMPSON: Give it a decade. Microsoft has short-term things and long-term things. This is one of the long-term things, like Windows.

METCALFE: At the risk of being nasty, what I just heard is that we need standardization. This is all I heard. I did not hear that all this money we are spending on software research is not resulting in any breakthroughs, or whatever breakthroughs it is resulting in are not being converted because we just cannot standardize on it. Is this right? Is this what I heard?

SHAW: Standardization suggests that there is one size that fits all, and if everyone would "just do it *my* way, everything would be just fine." That implies that there is one way that suffices for all problems.

LUCKY: Isn't standardization what made the Web? We all got together behind one solution; it may not fit everybody, but we empowered everybody to build on the same thing, and this is what made the whole thing happen.

CLARK: One statement that was made at the beginning of this decade was that the nineties would be the decade of standards. There is an old joke: the nice thing about standards is that there are so many to pick from. In truth, I think that one of the things that has happened in the nineties is that a few standards—not because they are necessarily best—happened to win some sort of battle.

LUCKY: This is a tragedy and a great triumph at the same time. You can build a better processor than Intel or a better operating system than Microsoft. It does not matter. It just does not matter.

CLARK: How can you hurtle into the future at a reckless pace and, simultaneously, conclude that it is all over, it does not matter because you cannot do something better, because it is all frozen in standards?

METCALFE: There seems to be reckless innovation on almost all fronts except two, software engineering and the telco monopolies.

CLARK: Yet if we look at the Web, the fact is that we have a regrettable set of de facto standards in HTML and HTTP, both of which any technologist would love to hate. When you try to innovate by saying it would be better if URLs were different, the answer is, "Yes, well there are only 50 million of them outstanding, so go away." Therefore, I am not sure I believe your statement that there is rapid innovation everywhere, except for these two areas.

METCALFE: I go back to Butler Lampson's comments. Just last week there was rapid innovation in the Web.

CLARK: How about Windows 95?

METCALFE: Windows 95 is an endgame.

LUCKY: Dave, it is possible that if all the dreams of the Java advocates come true, this will permit innovation on top of a standard. It is one way to get at this problem. We do not know how it is going to work out, but at least this would be the theory.

CLARK: I actually believe it might be true. I think this is very interesting.

A tremendous engine exists down below that is really driving the field—the rate of at least performance innovation, if not cost reduction, in the silicon industry. This was the engine that drove us forward. I think that this is true, but I am not sure it's the only engine. I wonder if on our 20th anniversary we will say, "Well, yes, silicon is the thing that drove us forward"; or will there be other things? Is the World Wide Web a creation of silicon innovation?

SHAW: No, it is a creation of the frustration of people who did not feel like dealing with FTP and TELNET but still wanted to get to information.

CLARK: I think you just said that silicon and frustration are our drivers.

LUCKY: At the base, silicon has driven the whole thing. It has really made everything possible. This is undeniable, even though we spend most of our time, all of us, working on a different level. This is the engine in the basement that really is doing it.

METCALFE: The old adage: "Grove giveth and Gates taketh away."

CLARK: You know I am an academic researcher. I thought I would ask the panel a question about my future, because I am very concerned about this. We heard all sorts of things earlier in the symposium about the nature of the field and the relationships that exist among research activities. I will be somewhat parochial here in order to focus. For academic research, there is the model of the future of reckless innovation, combined with the alternative model of well, it is all over. Somebody said to me that you can build a better operating system, but it would not matter. You can make a better Web, but it would not matter. You can create a better computer architecture, but it would not matter. It is all over.

In some places, like the silicon industry, I heard that the vector is very clear. They can see all the way out to 2010, and they know the problems they have to solve. I cannot repeat their language because I do not speak their language, but they have to learn how to do bipolar implant polarization. It is advanced technology development. In that context, Howard Frank considers what the research community is doing as very narrow.

I agree. If, in fact, our agenda has been defined by the boundary conditions of Windows 95 and the insistence of the silicon industry to move forward, then I think it is narrow, and there is no funding. If I had a good idea, I can bring one or two FTEs (full-time equivalents) to bear on it, and industry could bring 100 man-years to bear on it. Microsoft cranked out ActiveX in a year, right? How many man-years are in that? So what role can a poor academic play? I find myself asking, "If all of the academic researchers died, what impact would it have on the field in 10 years?"

REDDY: No students.

LUCKY: It is like the NBA (National Basketball Association) draft. Students are going to be leaving early, trying to be Mark Andreessen.

CLARK: This has happened to me. I cannot get them to stay. There is no doubt that it is a serious issue for me. So why does it matter?

METCALFE: I think it is true that, right now, industrial advancement in technology is outstripping the universities. I see this as a temporary problem that we need to fix. Some of us need to stop working on all these short-term projects in the universities and somehow leap out ahead of where the industry is now.

CLARK: I once described setting standards on the Internet as being chased by the four elephants of the apocalypse. The faster I ran, the faster they chased me because the only thing between them and making a billion dollars was the fact that you had not worked this thing out. You cannot outrun them. If it is a hardware area, you can hallucinate something so improbable you just cannot build it today. Then, of course, you cannot build it in the lab, either. We used to try to have hardware that let us live 10 years in the future. Now I am hard-pressed to get a PC on my desk. Yet in the software area, there really is no such thing as a long-term answer. If you can conceive it, somebody can reduce it to practice. So I do not know what it means to be long term anymore.

HARTMANIS: I do not believe what was said earlier, that if you invent a better operating system or a better Web or computer architecture, it does not matter. I think it matters a lot. It is not that industry takes over directly what you have done, but the students who move into industry take those ideas with them, and they do show up in development agendas and products. I am convinced that the above assessment is far too pessimistic about the influence of academic research.

FEIGENBAUM: I want to make a couple of comments. On the question of long-term versus short-term research, the universities would say, and I would say, that university researchers attend to longer-range issues. At a Defense Advanced Research Projects Agency (DARPA) software conference last August here in Washington, Bill Joy gave the keynote speech. In the question-and-answer period they talked about this issue of short-range and long-range research. In the context of stressing that there is really a place for universities, Bill said that at Sun, 18 months was a long time. He said he would not entertain anything that is more than 24 months out.

Then I was at another DARPA meeting recently where they were talking about advances in parallel computer architectures. The project they were focusing on as being very advanced was the work of the Stanford Computer System Lab on the FLASH architecture. This project has been going on for more than a decade now. It evolved with several different related architectures. This kind of sustained effort is the role of the university.

LUCKY: I want to say, in support of academics, that we are all proud of what the Internet and the Web have done. They were created by a partnership between academia and the government. The industry had very little to do with it. The question for all of us is whether this is a model that can be repeated. Can government do something again as it did with ARPANET that will have the tremendous effects for all of us that this has had two decades later?

WILLIAM WULF: I think this long-term versus short-term language is a red herring. Let me remind you of the figure that was up on the screen earlier this morning that comes from the Brooks-Sutherland report (Figure 2.1 in this volume). It shows research going on in universities and industry labs, product development going on, and the point in time at which something becomes a billion-dollar industry. The thing that is so wonderful about that figure is that it bounces back and forth. It is not a linear translation from far-out basic academic research to short-term grubbing product development. There are interesting, deep academic problems that are spawned by short-term product development. If any group in the world ought not to be having this discussion, it seems to me it is this field because we have experienced that the linear model is just so much nonsense.

MICHAEL DERTOUZOS: All this negativism, I just do not like it. Just a few random reflections: When I try to use my computer, I have to wait for what perceptually is 17 hours of booting. I do not want to wait that long, especially since I have to do it 20 or 30 times a day because it always fails. I would like a machine and a system that do not crash every 6 hours.

CLARK: Do you use Windows?

DERTOUZOS: I am using everything under the sun, and they all crash. I would like to have a machine that is easy to use. When I say a machine, I mean the whole spectrum—software and hardware systems. We brag about the Web, and yes, Butler, it is a great thing, and 40 million or 10 million users—or whatever the number is—is great, but there are 700 million telephones and 7 billion people in the world. There are voiceless millions, and we are not pinging against the limits. To get there and to have utility from these machines, we will have to be able to use them easily. To me, this is a long-term project for 30 or 40 years ahead.

Somebody said that voice was going away. I think speech is the most natural thing. We have to learn how to use it to make our machines understand us and learn from us. I just do not see this bottoming out of the field. Maybe we are in a bit of a lull. I agree that it is hard to find specific problems out there, but I think that if you look at the whole picture, there is a great deal ahead. I would like to ask if the people on the panel could provide their list of things that they would like to see.

FEIGENBAUM: I would like to say something about a paradox or a dilemma in which university researchers find themselves. If you go around and look at what individual faculty people do, you find smallish things in a world that seems to demand more team and system activity. There is not much money around to fund anything more than small things, basically to supplement a university professor's salary and a graduate student or two, and perhaps run them through the summer.

Partly this is because of a general lack of money. Partly it is because we have a population explosion problem and all these mouths to feed. All the agencies that were feeding relatively few mouths 20 years ago are now feeding maybe 100 times as many assistant professors and young researchers, so the amounts of money going to each are very small. This means that, except for the occasional brilliant meteor that comes through once in a while, you have relatively small things being done. When they get turned into anything, it is because the individual faculty member or student—as Professor Hartmanis mentioned, some students take these ideas out into the world—convinces an industry to spend more money on it. Subsequently, the world thinks that it came out of the industry.

ANITA BORG: I wanted to talk a little bit about the question of where you get innovation and where academics get ideas for problems to work on. This is something that I talk about every time I go, as an industry person, to talk to a university. It relates to what Bill was saying. If we keep training students to look inside their own heads and become professors, then we lose the path of innovation. If we train our students to look at what industry is doing and what customers and people out there using these things cannot do—not be terrorized by what they can do, but look at where they are running into walls—then our students start appreciating these as the sources of really hard problems. I think that this focus is lacking in academia to some extent and that looking outward at real problems gives you focus for research.

HARTMANIS: I fully agree. Students should be well aware of what industry is and is not doing, and I believe that many of them are well informed. Just as Michael Dertouzos complained about what is happening to his machine and what he wants to see done, students see problems with software and with the Internet. They go out and work summers in industry. They are not in any sense isolated; they know what is going on. Limited funding may not permit big university projects, but students are quite well informed about industrial activities.

SHAW: I side more with Anita. Earlier I mentioned three innovations that came from outside the computer science community—spreadsheets, text formatting, and the Web. I think they came about because people outside the community had something they needed to do and were not getting any help in doing. So we will get more leads by looking not only at the problems of computer scientists, but also at the problems of people who do not have the technical expertise to cope with these problems. I do not think the next innovation is particularly going to be an increment along the Web, or an increment on spreadsheets, or an increment on something else. What Anita is asking us to think about is, How are we going to be the originators of the next killer application, rather than waiting for somebody outside to show it to us?

FEIGENBAUM: I have talked to a lot of people abroad—academics and industry people in Japan and Europe—about our computer science situation, especially on the software side. We are the envy of the world in terms of the connectedness of our professors and our students to real-world problems. Talk about isolation—they think they are isolated relative to us.

I want to make a specific suggestion. There was a topic that came up in Joe Traub's talk about information warfare. There is, I think, a real-world context about which people ought to be concerned. I am giving you a perspective of 20 months with the Air Force and seeing the very real side of our academic discussions. We are in what I would call a pre-engineering phase with regard to handling the problems of information warfare that Joe spoke about. By pre-engineering I mean crafty and creative tinkering. If you actually go to the places where this work is done and watch the people at work, there is some scientific understanding, but not much. Indeed, there is no real engineering going on there, although the work is very innovative.

I think that the computer science academic world ought to pay attention. Len Kleinrock was making the case earlier that computer scientists and engineers should understand the nomadic computing world. He was telling us to understand this from the point of view of the "good guys" who want to give us functionality and ease of use. I would say we need to convince computer scientists and engineering researchers to understand the same world from the point of view of the "bad guys," and understand it at some depth. That is not the kind of thing that we academics usually pay attention to, but we must have good academic research focused on these issues.

STEWART PERSONICK: I want to add some data here. We have a research program at Bellcore; it is not enormous, but it amounts to about $35 million a year funded by our external customers, and we have some funding from the government as well. So we have a fair amount of money. In recent years, we have tried very hard to align our research to the needs of our customers to keep up the funding. I have advised universities that we funded at modest levels that I was not going to fund them as much as I used to. However, I indicated that I would be delighted to subcontract some of our research to them because this would be merely a transaction. I told them that it was not a gift. Bellcore has work to do, and we are prepared to subcontract it to them. We are talking potentially about millions of dollars. I have had no takers. People have been upset or discouraged by the fact that I have reduced the traditional funding; modest as it was, I have reduced it. I have not had anybody come back to me and express interest in subcontracting. What I hear people saying is, "Well, you know we don't do that."

This goes along with what I think Ed was saying. We do not have these enormous teams, but we do have teams working on big system problems that are very, very tough, and anyone who could solve these problems would be quite famous, in addition to making money for the customers. We are not seeing the academic community respond by saying it would love to subcontract this work. It seems as if it has not yet bought into this paradigm that we work together as a team on some problems that have real customers. These are not development things in some grubby sense. They are really, really tough computer science problems and system problems.

CLARK: Now it is time to give each of the panelists two or three minutes to tell us the thing about the future that matters the most to you.

REDDY: As Bob Lucky pointed out, there are different kinds of futures. If you go back 40 years, it was clear that certain things were going to have an impact on society—for example, communications satellites, predicted by Arthur Clarke; the invention of the computer; and the discovery of the DNA structure. At the same time, none of us had any idea of semiconductor memories or integrated circuits. We did not conceive of the ARPANET. All of these came to have an impact.

So my hypothesis is that there are some things we now know that will have impact. One is digital libraries. The term digital library is a misnomer, the wrong metaphor. It ought to be called digital archive, bookstore, and library. It provides access to information at some price, including no price. In fact, the National Science Foundation (NSF) and DARPA have large projects on digital libraries, but they are mainly technology based—creating that technology to access information. Nobody is working on the other problem of content.

We have a Library of Congress with 30 million volumes; globally, the estimate is about 100 million volumes. The U.S. Government Printing Office produces 40,000 documents consisting of 6 million pages that are out of copyright. Creating a movement—because it is not going to be done by any one country or any one group, it must be done globally—to get all the content (to use Jefferson's phrase, all the authored works of mankind) on-line is critically important. I think this is one of the futures that will affect every man, woman, and child, and we can do it. At Carnegie Mellon University (CMU), we are doing two things to help. In collaboration with National Academy Press, we are beginning to scan, convert, correct, and put in HTML format all of its out-of-print books. There are already about 200 to 300 of them. By the end of the year, we expect to have all of them. The second thing CMU is doing is offering to put all authored works of CSTB members on the network.

METCALFE: I would like speak briefly on behalf of efforts aimed at fixing the Internet. The Internet is one of our big success stories and we should be proud of it, but it is broken and on the verge of collapse. It is suffering numerous brownouts and outages. Increasingly, the people I talk to, numbering in the high 90 percent range now, are generally dissatisfied with the performance and reliability of the Internet.

There is no greater proof of this than the proliferation of intranets, which people tend to build. The good reason they build them is to serve internal corporate data processing applications, as they always have. The bad reason for building intranets is because the Internet offers inadequate security, performance, and reliability for its uses. So we now have a phenomenon in companies. The universities, as I understand it, are currently approaching NSF to build another NSFnet for them. This is really a suggestion not to fix the Internet, but to build another network for us.

Of course, the Internet service providers are also tempted to build their own copies of the Internet for special customers and so on. I believe that this is the wrong fix, the wrong approach. We need to be working on fixing the Internet. Lest you be in doubt about what this would include, it would mean adding facilities to the Internet by which it can be managed. I claim that these facilities are not in the Internet because universities find management boring and do not work on it. Fixing the Internet also would include the addition of mechanisms for finance so that the infrastructure can be grown through normal communications between supply and demand in our open markets, and the addition of security; it is not the National Security Agency's fault that we do not have security in the Internet. It occurred because for years and years working on security has been boring, and no one has been doing it; now we finally have started.

We need to add money to the Internet—not the finance part I just talked about, but electronic money that will support electronic commerce on the Internet. We need to introduce the concept of zoning in the Internet. The Communications Decency Act is an effort, although lame, to bring this about. On the Internet, mechanisms supporting freedom of speech have to be matched by mechanisms supporting freedom not to listen.

We need progress on the development of residential networking. The telecommunications monopolies have been in the way for 30 or 40 years, and we need to break these monopolies and get competition working on our behalf.

SHAW: I think the future is going to be shaped, as the past has been, by changes in the relationship between the people who use computing and the computing that they use. We have talked a lot today about software, and we have talked a little about the World Wide Web, which is really a provider of information rather than of computation at this point. I believe we should not think about these two things separately, but rather about their fusion as information services, including computation and information, but also the hybrid of active information.

On the Web, we have lots of information available as a vast undifferentiated sea of bits. We have some search engines that find us individual points. We need mechanisms that will allow us to search more systematically and to retain the context of the search. In order to fundamentally change the relation between the users and the computing, we need to find ways to make computing genuinely widespread and affordable, private and symmetric, and genuinely intellectually accessible by a wider collection of people.

I thank Bob Metcalfe for saying most of what I was going to say about what needs to be done because the networks must become places to do real business, rather than places to exchange information among friends. In addition, we need to spend more time thinking about what you might call naive models, that is, ways for people who are specialists in something other than computing to understand the computing medium and what it will do for them, and to do this in their own terms so they can take personal control over their computing.

LUCKY: There are two things I know about the future. First, after the turn of the century, one billion people will be using the Internet. The second thing I know is that I do not have the foggiest idea what they are going to be using it for.

REDDY: Digital libraries.

LUCKY: Perhaps. I think it is fundamental that we do not know this. We have created something much bigger than us where biological rules seem more relevant than the future paradigm that we are used to, where Darwinism and self-adaptive organization may be the more relevant phenomena with which to deal. The question is, How do we design an infrastructure in the face of this total unknown? There are certain things that seem to be an unalloyed good that we can strive for. One of them is bandwidth. Getting bandwidth out all the way to the user is something we can do without loss of generality.

On the other side, it is hard to find other unalloyed goods. For example, intelligence is not necessarily a good thing. Recently there was a flurry of e-mail on the Internet when one of the router companies announced that it was going to put an "Exon box" in its router. An Exon box would check all packets going by to see if they are adult packets or not. There was a lot of protest on the Internet, not because of First Amendment and Communications Decency Act principles, but because people did not want anything put inside the network that exercises control, simply as an architectural paradigm, more than anything else.

So it is hard to find these unalloyed goods. Bandwidth is good, but anything else you do on the network may later come back to bite you because of profound uncertainty about what is happening.

HARTMANIS: I would like to talk more about the science part of computer science, namely, theoretical work in computer science, its relevance, and identifying some stubborn intellectual problems. For example, security and trust on the Internet are of utmost importance; yet all the methods we use for encryption are based on unproven principles. We have no idea how hard it is to factor large integers, but our security systems are largely based on the assumed difficulty of factoring. There are many more such unresolved problems about the complexity of computations that are of direct relevance to trust, security, and authentication, as well as to the grand challenge of understanding what is and is not feasibly computable. The notorious $P = NP$ problem is probably the best known problem of this type, but by far not the only one. I consider these among the most important problems in theoretical computer science and sincerely hope that, during the next 10 years, some of them will be solved. I believe that deeper understanding of these problems will have a strong impact on computer science and beyond. Because of the universality of the computing paradigm, the quest to understand what is and is not feasibly computable is equivalent to understanding the limits of rational reasoning—a noble task indeed.

FEIGENBAUM: I would like to talk very briefly about artificial intelligence and the near future. If we look back 50 years—in fact to the very beginning of computing—Turing was around to give us a vision of artificial intelligence and what it would be, beautifully explicated in the play about Turing's life, *Breaking the Code*.

Raj Reddy published a paper in the May 1996 *Communications of the ACM*, his Turing Award address, called "To Dream the Possible Dream." I, too, share that possible dream. However, I feel like the character in the William Steig cartoon who is tumbling through space saying, "I hope to find out what it is all about before it is out."

There is a kind of Edisonian analogue to this. Yes, we have invented the light bulb, and we have given people plans to build the generators. We have given them tools for constructing the generators. They have gone out and hand-crafted a few generators. There is one lamppost working here, or lights on one city block are working over there. A few places are illuminated, but most of the world is still dark. Yet the dream is to light up the world! Edison, of course, invented an electric company. So the vision is to find out what it is we must do—and I am going to tell you what I think it is—and then go out and build that electric company.

What we learned over the past 25 years is that the driver of the power of intelligent systems is the knowledge the systems have about their universe of discourse, not the sophistication of the reasoning process the systems employ. We have put together tiny amounts of knowledge in very narrow, specialized areas in programs called *expert systems*. These are the individual lampposts or, at most, the city block. What we need built is a large, distributed knowledge base. The way to build it is the way the data space of the World Wide Web came about—a large number of individuals contributing their data to the nodes of the Web. In the case I am talking about, people will be contributing their knowledge in machine-usable form. The knowledge would be presented in a neutral and general way—a way of building knowledge bases so they are reusable and extendible—so that the knowledge can be used in many different applications. A lot of basic work has been done to enable this kind of infrastructure growth. I think we just need the will to go down that road.

DISCUSSION

JEROME GLENN: As far as being timid about talking about the future, didn't we all get into computers because we wanted to focus global intelligence on the most difficult problems to solve? Things that we could not do alone? Are we going to create a global interface between human brains and problems and machines? Isn't that the direction? So what is this fear of talking about the future?

PETER FREEMAN: I see a fair amount of confusion between the development of products or technology and the development of concepts or understanding. Several of you touched on this in your comments about what goes on, or should go on, in university-based research. I quite agree that many of us in universities are too focused on the short term, but ultimately, if we are to get to that next generation of products and technology, we have to have some new concepts.

I would point out just one in an area that several of you identified, software engineering, which I agree is almost devoid of ideas. There are a few people—Mary Shaw is one of them—who are trying to develop ways to express the architecture of software systems. Without that kind of architectural representation and description, we will never be able to do the kinds of things that Edward Feigenbaum was asking about, for example.

9

Unique Challenges: Computing and Telecommunications in a Knowledge Economy

Ellen M. Knapp

I will talk about discontinuities, opportunities, and challenges that we face as we move into the knowledge era, in the context of both the world economy and the business and organizational environment. I would like to highlight specifically some of the thoughts that I and others have had about the multifaceted role of computing and telecommunications in the transition from an industrial society to a knowledge society.

I want to be very clear that I am neither an academic researcher nor a corporate researcher. I am a corporate executive in the area of management. I am a vice chairman of Coopers & Lybrand, a $6 billion organization with approximately 70,000 partners and staff, operating in 146 countries at last count. As chief knowledge officer, I have responsibility for all aspects of technology as it relates to that enterprise and the interface of the enterprise, its suppliers, and its clients, as well as learning and education and market research and analysis.

Rather than bringing research results to this setting, I have created a knowledge-based collage. The remainder of my time is going to be a representation of a knowledge-based collage around the subject of computing and telecommunications in the knowledge economy.

There are a great many thoughts about this move from the industrial economy to the knowledge economy (Box 9.1). There is little argument about the fact that this transformation is taking place and that human capital and intellectual capital are going to be the principal sources of competitive advantage in this new era. There is little controversy about the shift in balance among labor, capital, land, and knowledge as we have gone from an agricultural to an industrial economy and now as we move toward a knowledge economy.

Representing the track that we are on into the knowledge era is not particularly controversial. However, it poses one very interesting question. That is, if you look at the relationship of 8,000 years to a century, roughly, or a little better than a century for the second major era, is the knowledge era going to be similarly collapsed with respect to its prior era? Or is it going to last for the next 8,000 years? There is also little controversy about what the principal sources of advantage are in these various eras (Figure 9.1). Where we begin to get into some interesting dialogue has to do not with the specific elapsed time of a particular era, but rather with the pace of the transformation or the transition from one era to the other. Different people have very different views of the pace (Box 9.2). Yet some people would point out that a very large part of the world's population is still living in the agricultural era. Not all of the world economy has even made the transition into the industrial era, even though that transformation took place in the middle of the last century.

There is even some controversy about whether or not we are in the knowledge era yet or whether we are entering the knowledge era, which means that it is somewhere slightly ahead of us. My view is more like the last

> **BOX 9.1 The Knowledge Economy**
>
> Knowledge is replacing matter and energy as the primary generator of wealth.
> —Thomas Stewart, Board of Editors, *Fortune*
>
> Today, knowledge and skills stand alone as the only source of comparative advantage.
> —Lester C. Thurow, *The Future of Capitalism*
>
> Tomorrow's economy will revolve around innovatively assembled brain power, not muscle power.
> —Tom Peters, *The Tom Peters Seminar*

line of Box 9.2, which says that the future has already turned into the present. Despite the fact that economists disagree on the pace of this transformation there is absolutely no argument about the fact that this transformation will have profound implications for all aspects of the world economy, the political scene, our individual personal and social behaviors, the environment, and the business environment.

There is, in fact, some concrete evidence that the knowledge era is already here. Some of that evidence is provided by both U.S. and global economic statistics. One of my favorites is this quotation from Thomas Stewart, Board of Editors of *Fortune*:

> 1991 was the crossover year when capital spending by U.S. companies was greater on telecommunications, copying and computer equipment than on industrial, construction, mining, and farming equipment.

We tend to think of the mid-1990s as the transition point, but we actually made this shift, from a macroeconomic perspective, around 1991.

Computing and telecommunications play very interesting roles with respect to this transition. After 25 years in the era of the computer, those roles are causing or catalyzing the transition from the industrial to the knowledge economy. In addition, computing and communications are also fundamental catalysts for the pace at which this transition is occurring. Unlike earlier transformations, the presence of and the advances within computing and telecommunications are, themselves, radically changing the pace of this transformation.

I recently had an opportunity to speak at an Massachusetts Institute of Technology faculty seminar series on this topic. I represented that one of the most intractable—but terribly interesting and incredibly important—problems we face right now is how to get a better handle on the pace of the transition from the industrial economy to the knowledge economy and of what it might look like. This is important for the following reason.

Economic Eras

Sources of Advantage

AGRICULTURAL ERA	INDUSTRIAL ERA	KNOWLEDGE ERA
Land	**Labor**	**Knowledge**
Natural Resources	**Engineering Skill**	**Skills**
Security	**Capital**	**Flexibility**
Population	**Market Access**	**Continuous Learning**

FIGURE 9.1 Economic eras and sources of advantage in each. SOURCE: Ron Bohlin, vice president, Digital Equipment Corporation, "Knowledge Networking: The Major Productivity Breakthrough."

> **BOX 9.2   Alternative Views of Timing of Transition Between Economic Eras**
>
> We have entered the knowledge economy.
> —Brooks Manville and Nathaniel Foote, McKinsey and Company
>
> Intellectual capital matters as we are leaving the industrial age and entering the information age.
> —Thomas Stewart, *Fortune*
>
> For several decades the world's best-known forecasters of societal change have predicted the emergence of a new economy in which brainpower, not machine power, is the critical resource. But the future has already turned into the present, and the era of knowledge has arrived.
> —"The Learning Organization," Economist Intelligence Unit

Many people have made the argument that world economies did not collapse when we went from the agricultural to the industrial era because there was a self-correcting mechanism during the transition phase. At the same time that labor was moving out of agricultural activities, there were new businesses, enterprises, and domains of human effort created in the industrial age that picked up this labor. As a result, there were no massive discontinuities in the work force. We do not know whether this will also be true of the next transition. We could, in fact, have some massive discontinuities in the work force in the United States and globally. We will not understand what is needed in the way of policy, and what is needed in the way of intervention to deal with those issues, unless we have a better understanding of the pace of change we are facing.

To understand the implications a little better, I would like to back up a bit and say a few words about the industrial era in order to encapsulate a few thoughts about the knowledge era. One of the implications of the numbers and the information in Box 9.3 is that during most of the industrial age companies had to be located somewhere. They had a natural home. If you look at the industries that the Japanese Ministry of International Trade and Industry (MITI) put in its *Vision for the Decade* in 1990 (Box 9.4), all of them are what Lester Thurow refers to in his latest book (*The Future of Capitalism*, 1996) as *brainpower industries*. One of the attributes of brainpower industries is that they do not have a natural home. These industries, and the winners in these industries, can reside anywhere that someone has the capability to mobilize the brainpower required to be a winner in them.

Today, economists (as well as the World Bank, which has published some statistics on this subject) estimate that human capital accounts for more than half of all of the wealth in the United States and other economically advanced nations. However, the difference between today and the future is that during the era of natural resources,

> **BOX 9.3   Aspects of the Industrial Era in Contrast with Those Anticipated in the Knowledge Era**
>
> Of the 12 largest industrial firms in the U.S. on January 1, 1990 . . .
> - 10 of the 12 were natural resource companies.
> - Only 1 of these companies—GE—is alive today.
>
> Implications
> - Those countries with natural resources were rich, and those without were destined to be poor.
> - For most of the industrial age, companies had natural homes where they had to be located.
>
> SOURCE: Lester Thurow, *The Future of Capitalism*.

> **BOX 9.4 Leading Industries and Implications in the Knowledge Era—MITI's View**
>
> In 1990, MITI published a list of industries it expected to be the most rapidly growing industries in the 1990s and into the early 21st century (*The Vision for the Decade*)
> - Microelectronics
> - Biotechnology
> - The new material sciences
> - Telecommunications
> - Civilian aircraft manufacturing
> - Machine tools and robotics
> - Computers (software and hardware)
>
> Implications
> - All of them are man-made brainpower industries.
> - All of them could be located anywhere on Earth—they have no natural home.
>
> SOURCE: Lester Thurow, *The Future of Capitalism*, 1996.

if you had natural resources you were destined to be rich; if you did not, you were destined to be poor. Just because human capital represents this aspect of wealth in our nation today does not necessarily imply that we are going to be a global winner in this knowledge era without deliberate intent.

I have some thoughts for the Computer Science and Telecommunications Board (CSTB) in terms of challenges. At a macroeconomic level, I think we must better understand and prepare for the profound changes in the world economy, U.S. economy, business environment, and social structures during "this period of punctuated equilibrium," as Lester Thurow calls it. For business organizations, which is the world I live in, I think the most interesting representation of what is going on in business is the quote from Paul Saffo at the Institute for the Future:

> It is hardly news that the corporation as we know it is headed for the scrap heap of business history. Internal corporate structures are already mutating beyond recognition. Corporate boundaries are dissolving into commercial irrelevance as businesses explore entirely new modes of association and interaction.

John Major, in his remarks at this symposium, made some very interesting and quite complementary comments about what is going on in the world of business and, even so, was referring primarily to the world of Motorola inside Motorola. I am going to talk a little about the inside of large multinationals like Motorola and my own organization, but equally interesting are the hypertext links being formed between multinationals and other enterprises.

Again, at the macroeconomic level, computing and telecommunications are playing a multifaceted role in this transformation of the business world. Figure 9.2 comes from *Business Week* (December 1993) and encapsulates some of the more unusual business models that were evolving as a result of the massive influx of computing and telecommunications into the business sector.

In work I did several years ago with Peter Keen (Figure 9.3), we came up with the notion of the relational business. That was before the Web was a major event. Now it is called the hypertext organization.[1] This is not quite the latest, but perhaps the second-to-latest, notion of what businesses are going to look like. As you can easily extrapolate, much of our ability to create organizations with this look and feel is empowered, if you will, by computing and telecommunications in our environments.

Equally interesting, the Institute for the Future recently came out with a publication that contains multiple models of types of twenty-first-century organizations. The one that clearly and most directly represents the

---

[1] Like an actual hypertext document, a hypertext organization is made up of interconnected layers or contexts: the business system, the project team, and the knowledge base. Nonaka, Ikujiro and Hirotaka Takeuchi. 1996. *The Knowledge-Creating Company: How Japanese Companies Create the Dynamics of Innovation*. Oxford University Press, Oxford.

Models for the Modern Organization

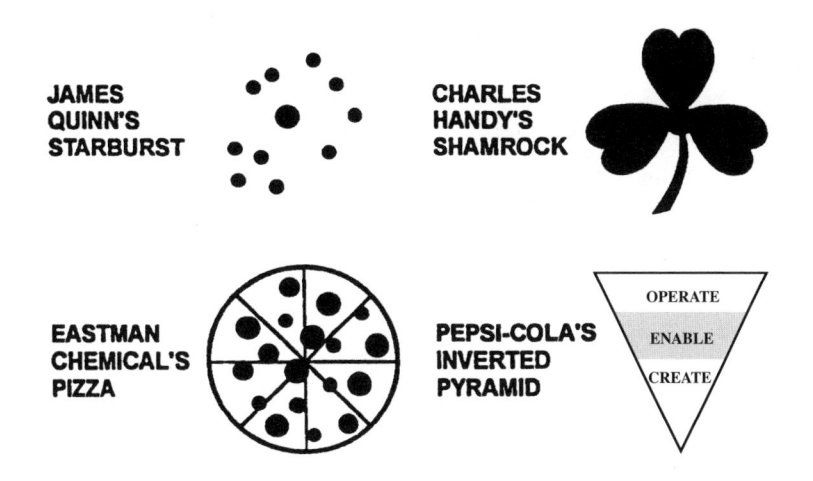

FIGURE 9.2 Business models for modern organization. SOURCE: "The Horizontal Corporation," *Business Week,* December 20, 1993.

industry that I am in, the professional services industry, is the fishnet model (Figure 9.4) and, in that vein—from a business strategy perspective—it is the only pure play in the knowledge economy. Professional services firms are knowledge-based, knowledge-centric, knowledge-intense organizations.

If anybody was going to have a chief knowledge officer (and I think it is a peculiar title myself) it would be an organization that has concluded that, in the future, both competitive advantage and comparative advantage will rest on its ability to mobilize both internal and external intellectual assets on behalf of its customers—and its ability to do it more efficiently and faster than its competitors. The fishnet organization does not incorporate notions of traditional secretaries, traditional flying around the world, or traditional buildings.

I will share with you a fact (not a particularly competitive one) from a strategy perspective. Some years ago—maybe five—I told our board that technology then represented the third largest cost element in our business model. I remind you that this is not a manufacturing or financial services firm, but a professional services firm. Even then, technology represented the third largest cost factor in our economic model. I told them that it would be third until it overtook facilities to become the second largest factor. Then it would be the second largest factor until it overtook human capital to become the largest. I said that the trend line was not going to change anytime in their lifetimes or mine.

The men in the boardroom looked at me as though I was crazy. Here we are, five years later, and technology has overtaken facilities as the second largest factor in our economic model. Just by going to a model in which—in this case, Motorola, BBN, DEC, and others—clients are providing the facility space for our people, as opposed to our organization providing it, the firm will substantially reduce costs.

Brian Quinn has also devised an organization model (*Harvard Business Review*, March-April 1996) that is a spider web (Figure 9.5). In a certain sense, whether it is a hypertext or a spider web or a relational or a fishnet model, they are all enabled, specifically and finitely, by the transformation that is taking place. I gave the example of Coopers & Lybrand adding millions of dollars to the bottom line just to show you that this is not all theoretical. Another very real example is occurring in the manufacturing sector, not the professional services sector.[2]

If it is true that long-term prosperity and comparative advantage, as well as competitive advantage, are going

---

[2] GE Lighting has closed 26 of 35 warehouses and replaced 25 customer service centers with one new, high-technology operation. Those buildings and stockpiles—physical assets—have been replaced by networks and databases—intellectual assets. Thomas Stewart, "Pursuing the Knowledge Advantage," *Fortune*.

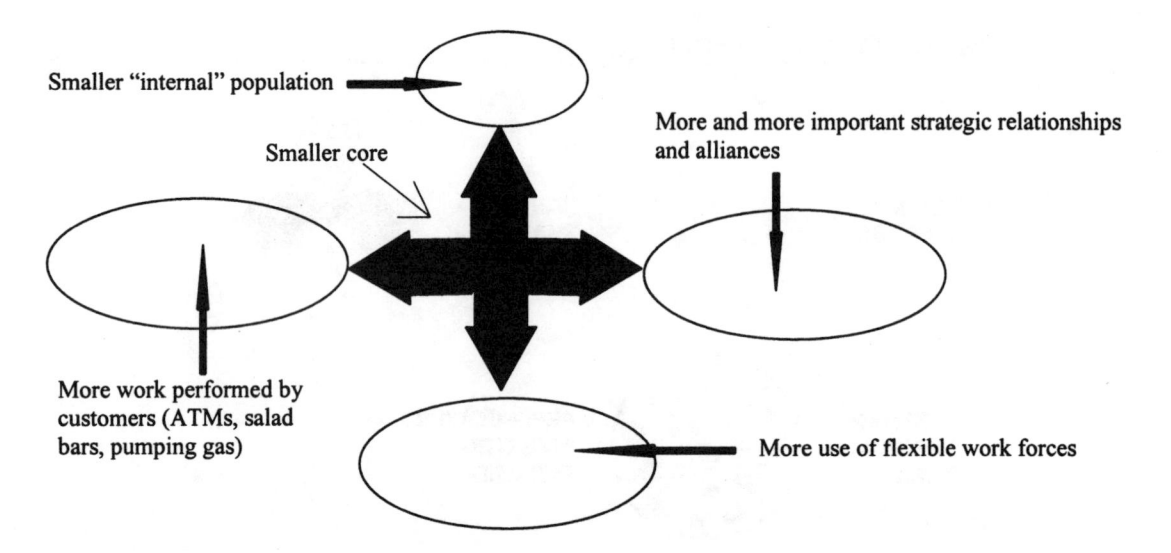

FIGURE 9.3 Relational model for the modern organization. SOURCE: Peter Keen, March 1995.

to come in large part from one organization's ability to manage its intellectual assets better than another, what is a model? What is a way of thinking about how a person might transform intellectual capital into revenue? Figure 9.6 is a very simple representation of what intellectual capital is—the transformation of human capital and structural capital into customer capital. The purpose, at least in our case because we happen to be a for-profit enterprise, is to create intellectual capital that can be transformed into financial capital. In my personal case, since I happen to be a partner, I want this transformed into wealth for shareholders.

Although this model represents a for-profit enterprise, it can also be used for not-for-profit enterprises. In fact, I have had several conversations about this model with the World Bank, and I will provide a short vignette to demonstrate that this is not a professional services phenomenon. It is an across-the-economy phenomenon.

The World Bank recently decided that it is not a bank, it never really was a bank, and it never should have

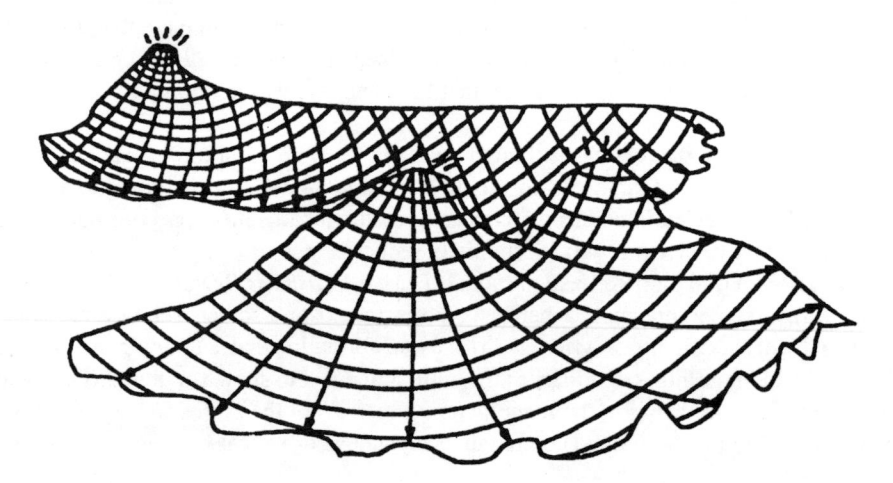

FIGURE 9.4 Fishnet organization model. SOURCE: Bob Johansen, Andrea Saveri, and Gregory Schmid, Institute for the Future, "21st Century Organizations: Reconciling Control and Empowerment."

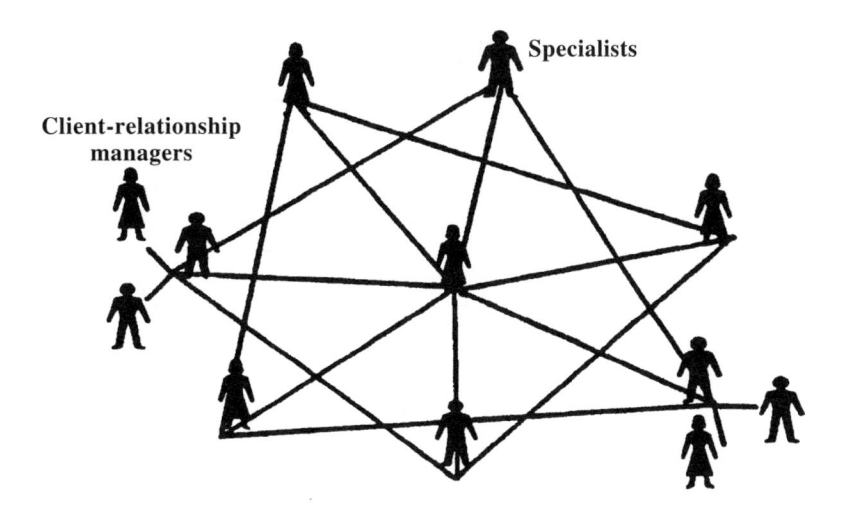

FIGURE 9.5 The spider web model for the modern organization. SOURCE: James Brian Quinn, Philip Anderson, and Sydney Finkelstein. 1996. "Managing Professional Intellect, " *Harvard Business Review,* March-April.

thought of itself as a bank. It is really a knowledge-centric professional services organization that, if it had only thought about this earlier, would have been able to generate (and still might in the future) enormous amounts of money—revenue is not quite the right term, but money, financial capital—through knowledge-based products and services. In fact, the World Bank has concluded that, about five years from now, its principal competitor will be Andersen Consulting. If the Bank wanted to look for best-practice, web organizations as models of what its computing and telecommunications infrastructure ought to look like, as well as organizational models, it ought to be looking at Andersen Consulting, not Citicorp. The Bank did not arrive at this notion all on its own, and I absolutely agree.

In this model, computing and telecommunications technology has two fundamentally different "plays." One play is in the domain of structural capital itself, and this is where we use technology to build organizational assets. We are building intranets, networks, intellectual capital, knowledge-based systems, and so forth—all for the purpose of building structural capital.

A very different role that computing and telecommunications play in this world is in the speed of the transformation process itself. Therefore, the ability to transform intellectual capital into financial capital and wealth—the speed with which competitors can go through this process—intrinsically is a major issue. At the same time we are moving into the information age in order to directly convert knowledge into revenue, we are also focused on creating and leveraging human capital and doing that with some speed and some facility.

The following quote by Andrew Grove, chief executive officer of Intel, that I have been carrying around with me for some years is still true today, and to be perfectly honest, I do not know how old it is.

> Computers will become communications platforms . . . people will use them to tie their work together, to collaborate . . . [in ways] that will revolutionize the way groups of people work.

We have moved fundamentally into a different world in terms of the reasons for which we expend financial capital on computing. Lest you conclude that I think that computing and telecommunications will create this whole new world all by themselves and propel us into the knowledge economy without attention to other details, I am not so naive. I understand that this will work only if people want it to work and are willing to make it work.

Our fundamental design challenge—for organizations at least, if not for individuals—revolves not around building better networks, but rather around building better *work*nets. If you want something interesting to get excited about, this is it.

Exploiting Technology for Knowledge Advantage

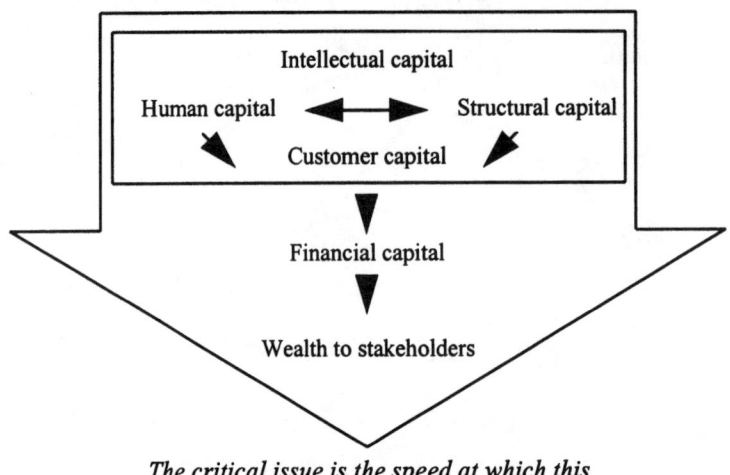

```
┌─────────────────────────────────────┐
│         Intellectual capital        │
│  Human capital  ⟷  Structural capital│
│         Customer capital            │
└─────────────────────────────────────┘
              │
              ▼
         Financial capital
              ▼
         Wealth to stakeholders
```

The critical issue is the speed at which this transforming process occurs.

FIGURE 9.6 An intellectual capital model. SOURCE: CIBC Leadership Center.

The collective power of computing and telecommunications can also bring us closer to our customers than we have ever been able to think about. Here is the good news and the bad news. Consider the Delta Shuttle card. There are only a few hundred of these in existence; I just happen to be the proud owner of one of them. Customers using the Delta Shuttle can put a credit card in a machine and get a ticket. Then you must go to the gate and get a boarding pass. You must keep your ticket so you do not lose it so that you can turn it in with your expense report. When I go to the Delta Shuttle, I go up to the gate, stick the Delta Shuttle card in a slot, and out comes not a ticket, but a boarding card. I take it and get on the plane. I have not spoken to a human being, and I do not need to carry any pieces of paper around with me or keep track of them. This phenomenon is most familiar to us with automatic teller machines. I also get out of my car and stick a card in the gas pump that I used to hand out the window to an attendant. This phenomenon has enormous social consequences, and we need to better understand it and its timing.

We face not only an issue of leveraging knowledge; we also have to manage knowledge. The sad fact is that, for the most part, we do not. We are still measuring the one thing that all of our executives learned something about when they were going through school, financial capital (Box 9.5). We have to get much better at looking at and understanding how to measure intellectual capital.

The conclusion is that if there is anybody in this room who is interested in the science of complexity or the emerging science of complex systems, this is one of the issues: the coevolution of economics, politics, social and environmental factors, and technology and what it is going to represent to our world is a very interesting, complex systems issue. Slightly more down to earth, I also have a couple of specific suggestions. We need a much better understanding of (1) the new business models, the organizational models, that are emerging; (2) what knowledge-centric cybercommunities are all about; (3) what we should think about in terms of continuous learning and delivery of continuous learning to highly dispersed populations; and (4) quality of life. We need a much better understanding of the pace of the massive shift in business process execution from labor to technology; the implications of this new era for wages, employment, and work location; intellectual capital asset management issues; and some of the new rules for the game that we are beginning to play.

So the bottom line is managing or, at best, figuring out how to adapt to the simultaneity. We know who the winners were when the issue was natural resources and the parameters of winning, the characteristics of how to win, when financial capital was the issue. We do not know who the winners are going to be in this knowledge era. We barely understand what the parameters of winning might look like.

> **BOX 9.5 Lack of Attention to Managing Knowledge Assets**
>
> Most corporations are still managed like old industrial companies . . . We're still managing our physical and financial assets, rather than our knowledge assets.
>   —Thomas Stewart, *Fortune*
>
> | How well do we manage assets? | | |
> |---|---|---|
> | Cash | A | Every penny at work all the time |
> | Receivables | A– | Well controlled, ratios ok |
> | Inventory | B+ | Mostly accurate |
> | People | C | People stuck in boxes |
> | Knowledge | D– | Huh?? Room for improvement? |
>
> SOURCE: Charles Savage, Knowledge Era Enterprises, Inc. "Implementing Knowledge Era Organizations."

DISCUSSION

DAVID MESSERSCHMITT: We have heard much about the significance of the fact that apparently the investment in telecommunications and computing has not increased productivity as measured in traditional terms. I get the sense that perhaps you do not believe this is the case. Perhaps more importantly, what I think I hear you saying is that it is missing the point. The increase in productivity in the industrial economy brought on by computers and telecommunications is not really the benefit; rather, it is creating a whole new economy. Could you comment on that?

ELLEN KNAPP: You are right. Unfortunately, you hit a very strong point with me. I happen to have been on the committee that produced the Computer Science and Telecommunication Board's report *Information Technology in the Service Society: A Twenty-First Century Lever*, and Stephen Roach at Morgan Stanley was also a member. It was really interesting, not the least of which is that he and I both have offices at 1251 Avenue of the Americas. We were in the same building and managed not to kill each other all those years until Marjory Blumenthal brought us together. To finish with the point that you ended with, I think that most of what Stephen had been measuring for many years was irrelevant. I think he was not asking the right questions about this transformation into a totally different world, rather than the productivity of a particular individual sitting behind a particular spreadsheet in the 1980s.

RAJ REDDY: This was a fascinating talk. There are a couple of points I wanted to make related to what you said. About seven or eight years ago, Edward Feigenbaum and I were talking about this issue of knowledge. If knowledge is truly the future wealth of nations, how could we demonstrate it? There were two ideas that came out of this. One is to create a world bank of knowledge. That is, take away all the money and only offer knowledge to people. The question is, Can they—taking the knowledge with all of the resources being available only for cost— create a completely new society out of nowhere? For example, if you built a wall around Washington, D.C., and the only thing that could come in and go out were knowledge, would you survive? It was an interesting question. What technology might exist? So the idea of creating a world knowledge bank came out of that. I thought this might be an interesting idea to talk about.

The second point you raised, that information technology is going to cause serious dislocations, is very important. Dislocations are going to happen much faster than society can handle. One thing I proposed a few years ago to an industry forum, which nearly lynched me, was that we should begin taxing the information industry—every computer and every piece of software at a 5 percent rate—and create a dislocation fund just like the North American Free Trade Agreement fund. Information technology is definitely going to cause dislocation. It is going to cause a lot of unhappiness in society. Middle managers of various kinds are going to be laid off, and a lot of other people, knowledge workers, are going to be laid off. Services industry people are going to be laid off. They need to be taken care of, reeducated, and retrained.

KNAPP: I will focus on the second point first. I believe that this is the first major world economic transformation that we have to address globally. Before we think about taxing information industries—which are predominantly in societies that did well in the industrial age—and potentially hampering the ability of that portion of the economy to uplift the other portions of the economy, I think we should look at the intricacies of the whole balancing act and think long and hard about it.

EDWARD FEIGENBAUM: What is the situation now with chief knowledge officers? How many companies have them? How have companies wrestled with the question of valuing knowledge assets for their assets and liability statement?

KNAPP: I have only been a chief knowledge officer since Monday. It was announced in the *Wall Street Journal*. I happened to be in Europe, so I did not get to read it, but I do not know a lot yet about who my colleagues are. I know that both McKinsey and Hewlett-Packard have chief knowledge officers. Coca-Cola has a chief learning officer, with roughly the same set of responsibilities. There are perhaps a half a dozen companies that have this title, not all in the services sector.

To address your second question, Skandia Corporation has, for the past three years, produced an intellectual asset balance sheet as a supplement to the annual report of the corporation. By far, Skandia and the Canadian Imperial Bank of Commerce are in the forefront of this field in terms of quantifying and measuring intellectual capital on an annual basis, and representing to the shareholder the fact that its valuation is as critical as financial capital valuation. Balance sheets that have to do with intellectual assets are as critical as balance sheets that have to do with financial capital assets. To my knowledge, Skandia is the only corporation today that produces these mirror-image balance sheets. I think you will see many, many more in the coming years.

MISCHA SCHWARTZ: You mentioned the need for a science of complexity. A previous speaker also alluded to some basic knowledge in systems complexity. It seems to me this may be something we cannot really do. Those of us who have been around a long time remember something called systems science that people worked on many years ago. The National Science Foundation had programs. I worked on something called "urban system analysis." It all fell apart. Even a system like the AT&T network fails occasionally. AT&T executives have been saying for years, "It is a shame we do not understand that complex system, simple as it is." Now you start putting people into it, is there really a possibility here?

KNAPP: There is a baby step that I think could be taken and has desperately been needed for many years. This is to get out of professional-academic, stovepipe-discipline problem solving. Even if we do not understand complexity theory and the science of complexity, we do understand team-based work. I do not know that universities do, but I do know that corporations do. If we could get some interdisciplinary, team-based work under way and energized within the university environment, I think that these teams could do some spectacular things that real-world people really need.

DAVID CLARK: I look at the Web, and one of the things the Web did was sort of "dis-intermediate" access to a wide variety of information. It may be information you think is valuable. Maybe it is not knowledge specifically. Maybe it is Joe Twiddle's home page, but at least one of the conclusions is that we have had a hard time selling this stuff for money. A lot of people are giving Web pages away free and hoping they can hide the advertising in there.

I am wondering if there is a possibility, in the long run, that we will succeed in making it possible for humans to have direct access not just to information, but to knowledge. At that point, you will not be able to sell it; the marginal cost of selling it is zero because there are not any humans in the loop. So there is not any value there. One possibility is that the whole market is doomed. Another one is that you think the product is the generation of knowledge and not the resale of it. Do you make most of your money creating knowledge or selling it multiple times? If you are going to sell it multiple times, do you think you can make any money at it, or do you have to sell it at zero marginal cost?

KNAPP: That is a very interesting question. We had an earlier speaker who passed on the question. I have some thoughts that I would be happy to share with you after the session.

10

Ancient Humans in the Information Age

Michael L. Dertouzos

We have brought upon ourselves and the world something that encompasses much more than the information society.[1] If one asks, "What is the value of information?" it quickly becomes apparent that the traditional theories of information are not effective in determining the value of a text editor, the work that an information worker does, a rented video, or an electronic form used in electronic commerce.

In the current, rapidly changing environment, people are confused about the nature of information. We hear that information is not scarce since it can be copied easily and therefore has very little value. Maybe it should be free, or maybe it should have a fixed cost. Part of the confusion stems from the fact that we view information mostly as a noun and forget that it is also a verb. As a noun, information consists of text, pictures, movies, and videos; as a verb, it refers to information-transforming work carried out either by a person such as a tax accountant or by a computer—for example, a word processing program.

Let me start with the assertion that, in economic terms, there is no difference between physical work and information work. Either one is produced by people who are reimbursed for expending a portion of their lives to do such work. Alternatively, the work is produced by computers that, like any other piece of equipment, require capital to be purchased. Thus, whether office work is produced by a human or a machine, it involves the same factors of production—labor and capital—as physical work.

If I project into the future, I envision a fairly simple model of what the information age is all about. I call it the *information marketplace.* It is a collection of people and their machines engaged in buying, selling, and freely exchanging information. It is a bit like an old village marketplace, except that what is exchanged is information rather than physical goods.

To address the value of information in this setting, let me divide economic goods and services into informational and physical, as well as intermediate and final. Final refers to something that is produced and then consumed, such as a loaf of bread. Intermediate refers to something used to produce a final or another intermediate good, such as flour. Final information encompasses items like books, entertainment, and videos—things that we consume for self-actualization, whose purpose ends there. When we add them up, these final goods comprise about 3 percent of the U.S. gross national product, leading us to the conclusion that the amount of final information today is very small.

---

[1]These brief comments are discussed in greater detail in the author's 1997 book, *What Will Be: How the New World of Information Will Change Our Lives,* Harper Edge, San Francisco.

Most information is intermediate because more than half of the work force is made up of office or knowledge workers. These employees and their machines are doing intermediate information work. Whereas final information is subject to the rules of supply and demand, intermediate information always goes toward enabling something else—eventually final information or, more often, physical final and intermediate products as well as services. Thus, the bulk of information has the important property of pointing to something else, leading to something, making something else possible.

The value of intermediate information is derived from what it points to. For example, at General Motors, all of the computers, software, and people working in the office represent the intermediate information that goes into making cars. The monetary value of all these activities is less than the value of the cars they sell and is derived from it. A huge amount of the U.S. economic basket is filled with physical goods and services. This means that there are many things for information to point to and derive value from. Employing computers and software makes a country more efficient and increases its wealth. In the United States, we value the hardware and software that point to these goods at 10 percent of the U.S. economy. On the other hand, in a poor country such as Bangladesh, the figure is less than 0.1 percent. This disparity illustrates that the rich countries (and people) value information much more than the poor simply because they have more economic goods to which information points. Also, since information technology helps those who use it to improve their productivity, we have an unstable situation in which the rich will get richer while the poor will stay behind. Left to its own devices, our technology is going to increase the gap between rich and poor. This calls for action and help on our part to ensure that it does not happen.

Let me now shift to what I call *electronic proximity*. Proximity and mobility are two sides of the same coin. The more mobile you are, the more people you can get close to. In the village age, we had about a couple of hundred neighbors whom we visited on foot, so our proximity was several hundred people. In the industrial age, cars increased our proximity by a thousandfold. We could drive a few hours and, potentially, reach hundreds of thousands of people. We did not have to know them all, but we could reach them. The information age will now give us another thousandfold increase—to hundreds of millions of people who will be within electronic reach. This is because we have 100 million computers connected today and (I forecast) will have some 500 million machines in five to seven years. This huge new increase in proximity is worrisome if we consider, by analogy, the problems that urban areas are facing today. We need to pay a great deal of attention to the problems of the forthcoming increased proximity.

When I think of proximity, I also think of telework. If people work from their homes, we are going to have a very interesting situation: a person will be an urban sophisticate by day, living in the world's markets; electronically commuting to Tokyo, Paris, and other major cities; carrying out all sorts of transactions, selling, buying, and fully exchanging knowledge and information. However, when the time comes to turn off the computer, the same individual will turn around and go out for pizza at a favorite local restaurant like a villager. She is an urbanite by day and a villager by night. We do not know which part of this split human will win the battle or even if both parts will learn to coexist within us.

Another point concerns nations and boundaries. Nations are located in one landmass because of their natural resources. France had wine. England and Germany had steel. We Greeks had grapes, knowledge, and democracy! Another reason nations remain physically compact is because they have a history, a set of traditions, and are protected by tribal unification forces. Now, the economic value of local resources has gone out the window, as the Japanese have shown the world. As for culture and history, consider this: In the future, I could dial up my high-speed network from Boston to Athens. I could sip ouzo while chatting with my friends in the Plaka, sing Greek songs, attend services at the Athens cathedral, or watch the sunrise on Santorini Island. I could partake of a lot of cultural, historic, and tribal "food" in a way that is not available to me in Boston today. So, all the old factors I mentioned that bind a nation within a common landmass seem to be disappearing. Perhaps tomorrow's Greece will not exist as a compact landmass, but as the Greek network!

Let me close with a third consideration, which is psychological. I submit that we are, today, the same ancient humans that Socrates and other more "normal" ancient people were in the past. We have the same body, mind, and psychology. Yet, our hands have moved from the stone club to the steering wheel, to the jet aircraft stick, and to the WIMP (windows, icons, menus, pointing) interface—a huge change. How are we coping with it? Bread-and-butter items such as text, photographs, videos, tables, and spreadsheets are fully transmitted over the information infrastructure. We are tempted to ask which activities and qualities actually pass through the airwaves, satellite links, and wires of the information infrastructure and which ones do not. It is clear that emotions are communicated to some degree. We all watch TV and sometimes laugh or cry at what we see and hear. However, emotions do not pass through fully, as pen pals can attest. Eventually, these virtual friends must consummate their relationship with a real-life experience such as meeting each other and shaking hands.

Are there some things that do not pass through the information infrastructure at all? When we lived in caves, we had a basic fear that animals would come from outside to eat our food or our children. This fear was a powerful force of the cave. Another force was when we hugged our loved ones and had physical contact with them. Just because ancient humans left the cave does not mean that these forces have left us. In fact, I suggest that these primal forces of the cave not only are still with us but are present for the most important, and even some of the mundane, decisions we make. Interactions with our siblings and friends; the relationships between doctors and patients; the trust between business associates or between students and teachers—these all involve the forces of the cave.

Do these human emotions pass through the wires and wireless links of the edifice we are constructing? I do not think so. For example, you can set up the best virtual reality full-immersion suit and create a robot designed to frighten me—like a monster from the cave—even hit me with its steel fist. I am wearing my body-net suit with virtual reality goggles and haptic gloves; I am seeing the monster approach, and it is getting very scary. However, I know that I can flip the switch and the monster will disappear. It is not a force of the cave unless it is real and I know it and feel it, not only rationally, but instinctively. Powerful and instinctive forces like the forces of the cave do not pass through the systems we are creating.

I conclude that the information age that lies ahead will not be a panacea, a paragon of knowledge, or a liberator of the human spirit as some of the current hype suggests. Instead, I believe that it will be a profound and powerful socioeconomic movement as big as the industrial revolution but ultimately of the same ilk, providing a new set of tools that will enable ancient humans like us to pursue our ancient goals and aspirations in new ways.

DISCUSSION

AUDIENCE PARTICIPANT: On this last point, Michael, I can remember in the fifties, back in the age of radio (which, by the way, was coming over wires or wireless), that this medium had a much greater capability to scare the living daylights out of me than TV or movies or Henry Fuchs's virtual reality because it was utilizing my own internalization of imagination and was deliberately limiting what was coming over the airways in that sense. So I am not sure I believe what you say, based on that experience.

MICHAEL DERTOUZOS: Oh, well, you are not scared as much as if a real force of the cave came after you. That would be my quick answer.

HENRY FUCHS: I agree. I am not going to scare anybody with virtual reality. I want to comment on your last, very wise observation. I want to take a slightly more pessimistic view. I will use the analogy of the telephone as a tool to let us, as you say, reach after the same forces of the cave. I think these new capabilities will allow us to extend the flexibility that the telephone has allowed us—to be physically in different places, but emotionally to still be together.

DERTOUZOS: One short comment, if I may. In business relationships, if you are trying to set up a serious merger or do something with a business partner, you will never do this by telephone unless you already know and have pressed the flesh of that person. I would say exactly the same thing.

FUCHS: Exactly right. So in this way, I suspect that what we will have is a situation in which you will not drink ouzo in Athens, in a small town, even with virtual reality, but there will be an added sense of sharing with people you already have a relationship with.

DERTOUZOS: Accepted. Bob, stump us.

ROBERT KAHN: Mike, you have described this flat-earth theory of physical work where, in essence, you consume the bread, you drop off the cliff, and it is really gone. You have got to do all the work again to create it one more time. You also presented a flat-earth theory of information. I wonder if the theory would not be better described as sort of the round-earth theory in the sense you said that, with the final information you get, you cannot figure out what you are going to do with it after that, so it must be final.

DERTOUZOS: Not always, Bob, but most of the time.

KAHN: Yet, in fact, you may not know what you are going to do with most things, and even if you thought you did, you might not. If final information was just that—information you did not know exactly what you were going to do with—then the whole educational system would be final information. I want to put forth the hypothesis that it is really a round-earth theory and that everything in the information world is really grist for somebody else.

DERTOUZOS: Well, you can close the loop back from education over a slower loop to the beginning of intermediate information. The point here is that you are not describing precise physical or human processes. You are developing a theory that has idealized components. When you have a real situation, it has pieces of them. I am doing this to try to understand what is going on with the economy.

EDWARD FEIGENBAUM: Mike, I wanted to take you up on this real life component versus the virtual reality component. I have to mention that Mike told me last night that Marvin Minsky had once said to him that if you gave humans perfect memory, you would not need more than one sexual experience. So how much real life do you think we have to mix in with the virtual life to make a workable modern life?

DERTOUZOS: I do not think I need to answer that. You reported precisely. What is good for Minsky, is good for the rest of the world.

APPENDIXES

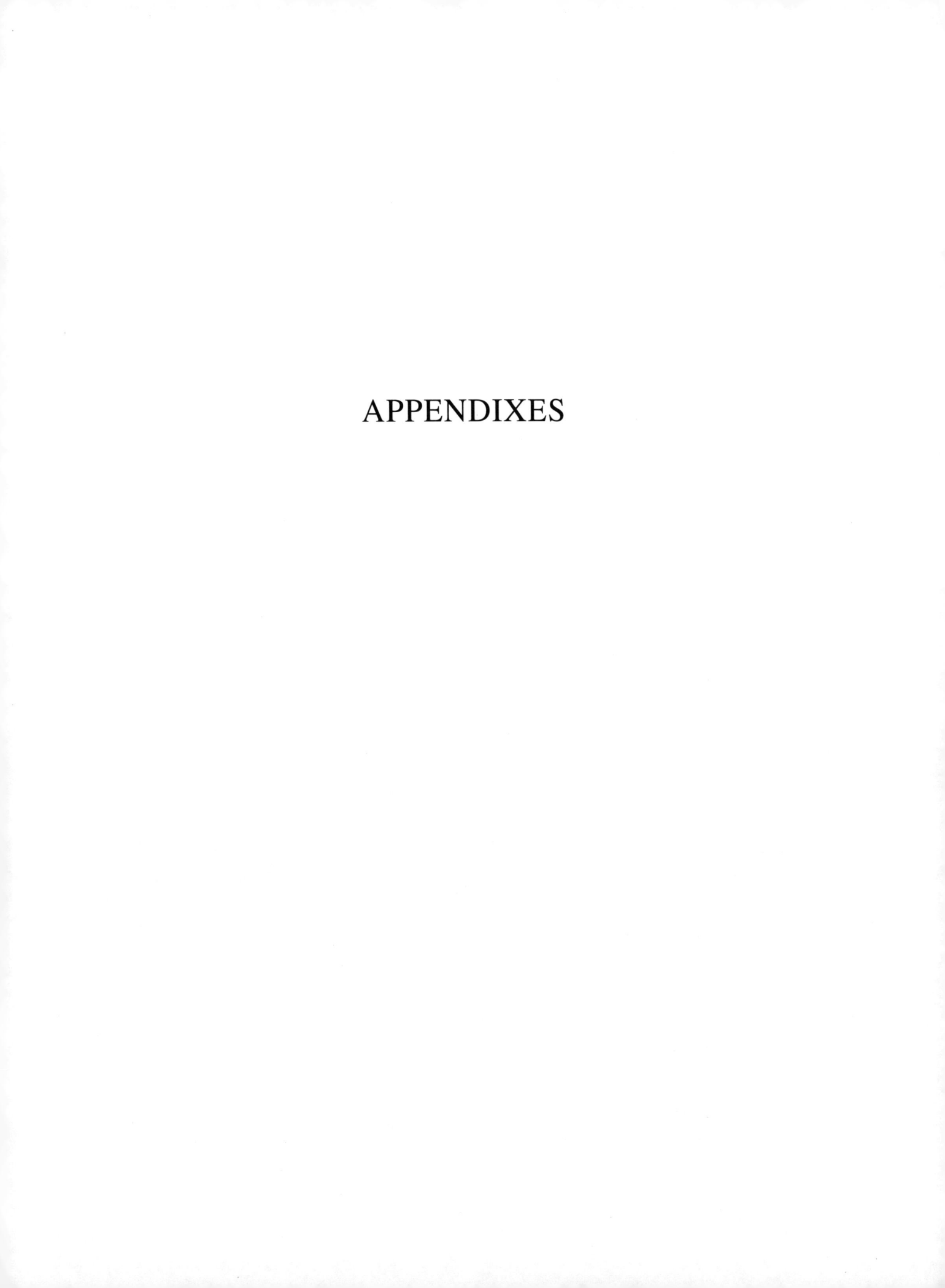

APPENDIX A

Letter from Dr. Bruce Alberts, President, National Academy of Sciences

May 1, 1996

Computer Science and Telecommunications Board
National Research Council
2101 Constitution Avenue, N.W.
Washington, DC 20418

Dear Members of the Computer Science and Telecommunications Board:

It gives me great pleasure to congratulate the Computer Science and Telecommunications Board on its 10th anniversary. I regret that other Academy commitments prevent me from welcoming you here today.

This symposium is a fitting way for the Board to celebrate 10 years of outstanding contributions to computer science, computing technology, and telecommunications. The National Academy of Sciences, National Academy of Engineering, and National Research Council are very proud of CSTB's achievements. The Board has acquired an outstanding reputation for its timely, relevant, and cogent work, completing more than 40 projects over the past decade. Several of its reports have a remarkably broad and enduring appeal. Personally, I have very much enjoyed reading *Realizing the Information Future* and *The Unpredictable Certainty*, and I have benefited greatly from some of the new knowledge that I have gained.

We live in an exciting, revolutionary time, and the next 10 years are likely to be even more productive for the Board and its staff. Computing and communications are changing the landscape dramatically in ways that affect every business, school, home, and citizen in the nation. CSTB and its expert committees will be challenged to respond to the needs of sponsors in search of answers to social and technological issues that could have broad ramifications for the public. In addition, CSTB must continue its tradition of generating its own project ideas that lead the nation in important directions.

On behalf of the National Academy of Sciences, National Academy of Engineering, and the National Research Council, I want to thank all former and current Board and committee members for their contributions to the growth and development of CSTB. In addition, accolades are due to the current Board Chair, Bill Wulf, and to the fine staff who have been essential to the Board's many accomplishments.

My congratulations to all associated with this fine endeavor. I hope that you enjoy this 10th Anniversary Symposium. Like me, I know that you look forward to another decade of exciting challenges for the Computer Science and Telecommunications Board.

Sincerely,

Bruce Alberts
President, National Academy of Sciences
Chairman, NRC Governing Board

APPENDIX B

Symposium Attendees

Marshall Abrams
The MITRE Corporation

Raymond Albers
Bell Atlantic Corporation

Frances Allen
IBM T.J. Watson Research Center

Saul Amarel
Rutgers University

Gary Anthes
Computerworld

Ruzena Bajcsy
University of Pennsylvania

Barbara Blaustein
National Science Foundation

Robert Bonometti
Bell Atlantic Corporation

Anita Borg
Digital Equipment Corporation

George Brandon
Telecomunications Report

Timothy Brennan
University of Maryland

Charles Brownstein
Cross-Industry Working Team

Aubrey Bush
National Science Foundation

Stephen Caimi
Citicorp Technology Office

Virginia Castor
Office of the Secretary of Defense

John Cavallini
U.S. Department of Energy

Jane Caviness
Educom

John Cherniavsky
National Science Foundation

Melvyn Ciment
National Science Foundation

David Clark
Massachusetts Institute of Technology

Clement Cole
Locus Computing Corporation

Eileen Collins
National Science Foundation

George Cotter
National Security Agency

Dorothy Denning
Georgetown University

Michael Dertouzos
Massachusetts Institute of Technology

Robert Direnzo
The World Bank

Jeff Dozier
University of California at Santa Barbara

Larry Druffel
South Carolina Research Association

David Farber
University of Pennsylvania

Edward Feigenbaum
U.S. Air Force

Jeanne Ferris
Chronicle of Higher Education

Kevin Finneran
Issues in Science and Technology

Francis Dummer Fisher
University of Texas at Austin

Barbara Fossum
IC2 Institute at the University of
  Texas at Austin

Howard Frank
Defense Advanced Research Projects Agency

Darleen Fisher
National Science Foundation

Peter Freeman
Georgia Institute of Technology

Henry Fuchs
University of North Carolina

Samuel Fuller
Digital Equipment Corporation

Christina Gabriel
National Science Foundation

Charles Geschke
Adobe Systems Incorporated

Helen Gigley
Office of Naval Research

Jerome Glenn
New Earth Radio

Norman Glick
National Security Agency

Seymour Goodman
University of Arizona

Ian Graig
Global Policy Group, Incorporated

Donald Greenberg
Cornell University

William Griffin
GTE Laboratories, Incorporated

Barbara Grosz
Harvard University

Juris Hartmanis
Cornell University

Donald Heath
The Internet Society

Harry Hedges
National Science Foundation

Carol Henderson
American Library Association

Lance Hoffman
The George Washington University

Lee Holcomb
National Aeronautics and Space Administration

Charles Holland
Air Force Office of Scientific Research

John Hopcroft
Cornell University

Mark Jacobs
Air Force Office of Scientific Research

Anita Jones
Department of Defense

Deborah Joseph
University of Wisconsin

Charles Judice
Eastman Kodak

Robert Kahn
Corporation for National Research Initiatives

Sidney Karin
University of California at San Diego

Julius Katz
Hills and Company

Stuart Katzke
National Institute of Standards and Technology

Stephen Thomas Kent
BBN Communications Corporation

Pradeep Khosla
Defense Advanced Research Projects Agency

Richard Kieburtz
Oregon Graduate Institute

Edwin Kiester
Reader's Digest

Sally Kiester
Reader's Digest

Thomas Kitchens
U.S. Department of Energy

Leonard Kleinrock
University of California at Los Angeles

Ellen Knapp
Coopers & Lybrand

William Kneisly
Thinking Machines Corporation

Butler Lampson
Microsoft Corporation

Neal Laurance
Ford Motor Company

Alfred Lee
U.S. Department of Commerce/NTIA

Donald Lindberg
The National Library of Medicine

Barbara Liskov
Massachusetts Institute of Technology

Allen Locke
Department of State

William Loveless
Federal Technology Report

Gloria Lubkin
Physics Today

Robert Lucky
Bellcore

Patrice Lyons
Attorney at Law

John Major
Motorola, Incorporated

Pamela McCorduck
Author

Francis McDonough
General Services Administration/ITS

David Messerschmitt
University of California at Berkeley

Robert Metcalfe
International Data Group

Melvin Montemerlo
National Aeronautics and Space Administration

Lisa Mount
Department of State

David Nelson
U.S. Department of Energy

Michael Nelson
Office of Science and Technology Policy

Susan Nunnery
Myricom, Incorporated

Rod Oldehoeft
U.S. Department of Energy

Stewart Personick
Bellcore

William Powless
Inside Energy/The McGraw Hill Companies

Arati Prabhakar
National Institute of Standards and Technology

William Press
Harvard College - Observatory

Raj Reddy
Carnegie Mellon University

John Phillip Riganati
David L. Sarnoff Research Center

Damian Saccocio
America OnLine

Paul Schneck
Mitretek Systems, Incorporated

Mischa Schwartz
Columbia University

Mary Anne Scott
Department of Energy

Charles Seitz
Myricom, Incorporated

Mary Shaw
Carnegie Mellon University

Edward Shortliffe
Stanford University School of Medicine

Margaret Simmons
National Coordination Office for High
    Performance Computing and Communications

Michael Simmon
Santa Fe Institute

Alexander Singer
Film Director

Judith Singer
Writer

Irwin Sitkin
Aetna (retired)

Paul Smith
National Coordination Office for High
    Performance Computing and Communications

Lawrence Snyder
University of Washington

Rowan Snyder
Coopers & Lybrand

Avi Spector
George Washington University

Hillary (Traub) Spector
Hillary's Fine Jewelry

William Spencer
SEMATECH

Robert Spinrad
Xerox Corporation

APPENDIX B

Gary Strong
National Science Foundation

Robert Sullivan
University of Texas

Lawrence Tesler
Apple Computer, Incorporated

Joseph Traub
Columbia University

Leslie Vadasz
Intel Corporation

Andries van Dam
Brown University

Andre van Tilborg
Office of Naval Research

Shukri Wakid
National Institute of Standards and Technology

Caroline Wardle
National Science Foundation

Jane Williams
U.S. National Commission on Libraries & Information Science

Shmuel Winograd
IBM T.J. Watson Research Center

Joan Winston
Trusted Information Systems, Incorporated

Fred Wood
National Library of Medicine

William Wulf
University of Virginia

David Wye
Federal Communications Commission

Wu Yishan
Embassy of China

Paul Young
National Science Foundation

APPENDIX C

Biographies of Presenters

DAVID D. CLARK, Massachusetts Institute of Technology

David Clark, who became CSTB's chair in 1996, graduated from Swarthmore College in 1966, and received his Ph.D. from MIT in 1973. He has worked since then at the MIT Laboratory for Computer Science, where he is currently a senior research scientist in charge of the Advanced Network Architecture group. Dr. Clark's research interests include networks, network protocols, operating systems, distributed systems, and computer and communications security. After receiving his Ph.D., he worked on the early stages of the ARPANET and on the development of token ring local area network technology. Since the mid-1970s, Dr. Clark has been involved in the development of the Internet. From 1981-1989, he acted as chief protocol architect in this development, and chaired the Internet Activities Board. His current research area is protocols and architectures for very large and very high-speed networks. Specific activities include extensions to the Internet to support real-time traffic, explicit allocation of service, pricing, and new network technologies. In the security area, Dr. Clark participated in the early development of the multi-level secure Multics operating system. He developed an information security model that stresses integrity of data rather than disclosure control. Dr. Clark is a member of the IEEE, the ACM, and the National Academy of Engineering. He received the ACM SIGCOMM award and the IEEE award in International Communications for his work on the Internet. He is a consultant to a number of companies and serves on the boards of two corporations. Dr. Clark chaired the committee that produced the CSTB report, *Computers at Risk: Safe Computing in the Information Age*. He also served on the committees that produced the CSTB reports, *Toward a National Research Network; Realizing the Information Future: The Internet and Beyond*; and *The Unpredictable Certainty: Information Infrastructure Through 2000*.

MICHAEL L. DERTOUZOS, Massachusetts Institute of Technology

Michael Dertouzos, a founding member of CSTB, is professor of computer science and electrical engineering and director of the MIT Laboratory for Computer Science—the home base of the World Wide Web. He is author or co-author of six books. One, *Made in America: Regaining the Productive Edge*, is the result of the study by the MIT Commission on Industrial Productivity, which he chaired. Dr. Dertouzos is a member of the National Academy of Engineering and an advisor to the U.S. and European governments. He concentrates his current efforts on the architecture, uses, and impact of tomorrow's information infrastructures. He accompanied Vice

President Albert Gore as a U.S. delegate to the G7 meeting on the Global Information Society in Brussels. Professor Dertouzos is a dual national of the United States and the European Union, and is involved on both sides of the Atlantic with the strategic steering of governments and large organizations into the Information Age. Dr. Dertouzos was the prime mover behind CSTB's report, *The National Challenge in Computer Science and Technology*.

EDWARD A. FEIGENBAUM, U.S. Air Force

Edward Feigenbaum, a founding member of CSTB, is chief scientist of the U.S. Air Force, Washington, D.C. He serves as chief scientific advisor to the chief of staff and the secretary and provides assessments on a wide range of scientific and technical issues affecting the Air Force mission. Dr. Feigenbaum has served on the DARPA Information Science and Technology study committee. He has been a faculty member at Stanford University for 29 years, and is founder and co-director of the Knowledge Systems Laboratory, a leading laboratory for work in knowledge engineering and expert systems. A professor of computer science, Dr. Feigenbaum is internationally known for his work in artificial intelligence and expert systems, and is the co-author of seven books and monographs as well as some 60 scientific papers. He is a co-founder of three start-up firms in applied artificial intelligence and has served on the board of directors of several companies. He has served as a member of the Board of Regents of the National Library of Medicine and as a member of NSF's Computer Science Advisory Board. Dr. Feigenbaum received his B.S. in electrical engineering from the Carnegie Institute of Technology, Pennsylvania; and his Ph.D. degree from the Graduate School of Industrial Administration, Carnegie Institute of Technology, Pennsylvania. In 1991, he was elected a member of the American Academy of Arts and Sciences and also received the Career Achievement Award from the World Congress on Expert Systems (the Feigenbaum Medal, named in his honor). He was elected a fellow of the American Institute of Medical and Biological Engineering in 1994, and he is a member of the National Academy of Engineering.

HOWARD FRANK, Defense Advanced Research Projects Agency

Howard Frank at the time of the symposium was director of DARPA's Information Technology Office, where he managed a $350-million annual budget aimed at advancing the frontiers of information technology. Dr. Frank was responsible for DARPA's research in advanced computing, communications, software, and intelligent systems, with programs ranging from language systems and human-computer interaction to scalable high-performance computing, networking, security, and microsystems. Before becoming director of ITO, he was the director of the Computing Systems Technology Office (now part of ITO), and earlier, special assistant to the director of DARPA for Information Infrastructure Technology. While at DARPA, Dr. Frank helped found the DARPA/DISA Joint Program Office, a joint activity with the Defense Information Systems Agency. He then completed a two-year assignment as its first director. In September 1997 Dr. Frank became the dean of the College of Business and Management at the University of Maryland.

Dr. Frank was chair of the Technology Policy Working Group (TPWG) of the administration's Information Infrastructure Task Force and led the TPWG's Advanced Digital Video and Security Process Projects. He is also DARPA's representative on the White House National Science and Technology Council's Committee on Information and Communications. Dr. Frank has been a member of six editorial boards and a featured speaker at over 100 business and professional meetings; he has authored over 190 articles and chapters in books. Dr. Frank is a fellow of the IEEE. He is a senior fellow at the Wharton School's SEI Center for Advanced Studies in Management. Before joining DARPA, Dr. Frank was an advisor to large companies in information and corporate strategy, market positioning, and mergers and acquisitions. Earlier, he was founder, chairman, and CEO of Network Management Inc.; president and CEO of Contel Information Systems (a subsidiary of Contel); president, CEO, and founder of the Network Analysis Corporation; a visiting consultant within the Executive Office of the President of the United States in charge of its network analysis activities; and an associate professor at the University of California, Berkeley. He served on the committee that produced the CSTB report, *Toward a National Research Network*.

HENRY FUCHS, University of North Carolina

Henry Fuchs, a CSTB member at the time of the symposium, is Federico Gil professor of computer science and adjunct professor of radiation oncology at the University of North Carolina at Chapel Hill. He received a B.A. in information and computer science from the University of California at Santa Cruz in 1970, and a Ph.D. in computer science from the University of Utah in 1975. He received the 1992 Computer Graphics Achievement Award from ACM/SIGGRAPH and the 1992 National Computer Graphics Association Academic Award. He was an associate editor of *ACM Transactions on Graphics* (1983-1988) and guest editor of its first issue (January 1982). He was the technical program chair for the ACM/SIGGRAPH 1981 Conference, chairman of the 1985 Chapel Hill Conference on Advanced Research in VLSI, chairman of the 1986 Chapel Hill Workshop on Interactive 3D Graphics, co-director of the NATO Advanced Research Workshop on 3D Imaging in Medicine (1990), and co-chair of the National Science Foundation Workshop on Research Directions in Virtual Environments (1992). He serves on various advisory committees for government and industrial groups.

SEYMOUR E. GOODMAN, University of Arizona

Seymour Goodman is professor of MIS and policy, a member of the Center for Middle Eastern Studies at the University of Arizona (since 1981), and Carnegie Science Fellow at the Center for International Security and Arms Control at Stanford University (since 1994). He studies the global diffusion and other international developments in information technologies and related public policy questions. Professor Goodman has had various permanent and visiting appointments at the University of Virginia (computer science, Center for Soviet and East European Studies), Princeton University (mathematics, Woodrow Wilson School of Public and International Affairs), and the University of Chicago (economics). He has served on numerous government, academic, and industry study and advisory committees, and is contributing editor for international perspectives for *The Communications of the ACM*. He has visited all seven continents and approximately 70 countries during the past 15 years. Professor Goodman was an undergraduate at Columbia University and received his Ph.D. from the California Institute of Technology. He chaired the committee that produced CSTB's report, *Global Trends in Computer Technology and Their Impact on Export Control.*

DONALD P. GREENBERG, Cornell University

Donald Greenberg is the Jacob Gould Schurman professor of computer graphics at Cornell University. He was the founding director of NSF's Science and Technology Center for Computer Graphics and Scientific Visualization. He is also the director of the Program of Computer Graphics and former director of the Computer-Aided Design Instructional Facility at Cornell University. Since 1965, he has been researching and teaching in the field of computer graphics; he is primarily concerned with physically-based image synthesis and with applying graphic techniques to a variety of disciplines. His specialties include color science, parallel processing, and realistic image generation. He teaches computer graphics, computer-aided design, digital photography, and computer information courses in the computer science, architecture, art, and business schools, respectively. His application work now focuses on medical imaging, architectural design, digital photography, and interactive video. In 1987, he received the ACM Steven Coons Award, the highest honor in the field, for his outstanding creative contributions in computer graphics. He also received the National Computer Graphics Association Academic Award in 1989. He is a member of the National Academy of Engineering and a fellow of the International Association of Medical and Biological Engineering and of the ACM.

JURIS HARTMANIS, Cornell University

Juris Hartmanis at the time of the symposium was a CSTB member and the Walter R. Read professor of engineering and computer science at Cornell University. In 1996 he became NSF's Assistant Director for Computer and Information Science and Engineering. He was the first chair of Cornell's computer science department, founded in 1965, and had been at Cornell since then. His research interests include theory of computation and computational complexity. He shared the 1993 ACM Turing Award with R.E. Stearns for their seminal work on computational complexity. In 1995, he received the B. Bolzan Gold Medal from the Academy of Sciences of the Czech Republic and the honorary Dr.h.c. from the University of Dortmund, Germany. Dr. Hartmanis is a member of the National Academy of Engineering, American Academy of Arts and Sciences, New York Academy of Sciences, fellow of the ACM and AAAS, and foreign member of the Latvian Academy of Sciences. He has authored two books, *Algebraic Structure Theory of Sequential Machines* and *Feasible Computations and Provable Complexity Properties*. In addition, he has authored over 140 research papers. Dr. Hartmanis graduated with a B.S. in physics from the University of Marburg in 1949. He received his M.A. in mathematics from the University of Kansas City and his Ph.D. in mathematics from the California Institute of Technology in 1951 and 1955, respectively. He chaired the committee that produced the CSTB report, *Computing the Future: A Broader Agenda for Computer Science and Engineering*.

DEBORAH A. JOSEPH, University of Wisconsin

Deborah Joseph, a current CSTB member, is an associate professor of computer science and mathematics at the University of Wisconsin. She received a B.A. (interdisciplinary-ecology, 1976) from Hiram College, and an M.S. (computer science, 1978) and a Ph.D. (computer science, 1981), both from Purdue University. Dr. Joseph held the National Science Foundation's Presidential Young Investigator Award for 1985-1990. Her research interests include complexity theory, computational problems in molecular biology, computational geometry, and mathematical logic-recursion theory.

LEONARD KLEINROCK, University of California, Los Angeles

Leonard Kleinrock, a founding member of CSTB, has been a professor of computer science at the University of California, Los Angeles, since 1963. He received his B.S. degree in electrical engineering from the City College of New York in 1957, and his M.S. and Ph.D. degrees in electrical engineering from MIT in 1959 and 1963, respectively. His research interests focus on performance evaluation of high-speed networks and parallel and distributed systems. He has had over 190 papers published and is the author of five books. He is the principal investigator for the DARPA Advanced Networking and Distributed Systems grant at UCLA. He is also founder and CEO of Technology Transfer Institute, a computer-communications seminar and consulting organization located in Santa Monica, CA.

Dr. Kleinrock is a member of the National Academy of Engineering, a Guggenheim Fellow, and an IEEE fellow. He has received numerous best paper and teaching awards, including the ICC 1978 Prize Winning Paper Award, the 1976 Lanchester Prize for outstanding work in operations research, and the Communications Society 1975 Leonard G. Abraham Prize Paper Award. In 1982, he received the Townsend Harris Medal. Also in 1982, he was co-winner of the L. M. Ericsson Prize, presented by His Majesty King Carl Gustaf of Sweden, for his outstanding contribution in packet switching technology. In July 1986, Dr. Kleinrock received the 12th Marconi International Fellowship Award, presented by His Royal Highness Prince Albert, brother of King Baudouin of Belgium, for his pioneering work in the field of computer networks. In the same year, he received the UCLA Outstanding Teacher Award. In 1990, he received the ACM SIGCOMM award recognizing his seminal role in developing methods for analyzing packet network technology, and in 1996, he was given the Harry Goode Award. Dr. Kleinrock chaired the committees that produced the CSTB reports, *Toward a National Research Network* and *Realizing the Information Future: The Internet and Beyond*. He served on the committee that produced the CSTB report, *Computing the Future: A Broader Agenda for Computer Science and Engineering*.

ELLEN M. KNAPP, Coopers & Lybrand

Ellen Knapp is vice chairman, technology, for Coopers & Lybrand, L.L.P., and a member of the firm's management committee and its board of partners. She is responsible for providing technology leadership and embedding technology in all lines of business within the firm. In addition, Ms. Knapp is chairman of C&L's International Technology Management Group, providing strategy, policy, and standards to C&L member firms, and ensuring consistency of service to C&L's multi-national clients. C&L International, through its member firms, operates in 750 offices in 140 countries with a 1995 revenue of $5.8 billion. Ms. Knapp is responsible for developing pioneering uses of technology throughout all areas of the firm, including quality and efficiency of client services, internal operations, and creation of innovative services to clients. One of the country's primary authorities on the strategic use of technology, Knapp's role is to create competitive advantage and catalyze organizational change in C&L through the development and application of leading-edge technology. Formerly, Ms. Knapp was a senior partner in Coopers & Lybrand's management consulting services practice and national director of information technology consulting for the U.S. firm.

Before joining C&L, Ms. Knapp was responsible for establishing the advanced technology consulting practice at Booz-Allen & Hamilton. She has provided both technical and management consulting services to a wide range of public and private sector clients, including transnational consumer products companies, telecommunications firms, large service sector organizations, and worldwide manufacturing enterprises. Ms. Knapp has received worldwide recognition for her work in advanced technologies. She has published professional papers, contributed to edited texts, and served as a speaker or moderator at numerous symposia, television appearances, and press conferences in the United States, Europe, and Asia. She is a featured speaker at the 1996 *Women Shaping Technology* conference and a juror for the 1996 Lemelson-MIT Award for Invention and Innovation. She has co-authored two books with Peter Keen, *Every Manager's Guide to Business Processes* (1995) and *Process Payoffs: Building Value Through Business Process Investment* (in publication), both published by Harvard Business School Press. She served on the committee that produced the CSTB report, *Information Technology in the Service Society: A Twenty-First Century Lever.*

ROBERT W. LUCKY, Bellcore

Robert Lucky, a founding member of CSTB, is corporate vice president of applied research at Bellcore. Born in Pittsburgh, PA, he attended Purdue University, where he received a B.S. degree in electrical engineering in 1957, and M.S. and Ph.D. degrees in 1959 and 1961. After graduation, he joined AT&T Bell Laboratories in Holmdel, NJ, where he was initially involved in studying ways of sending digital information over telephone lines. The best known outcome of this work was his invention of the adaptive equalizer—a technique for correcting distortion in telephone signals that is used in all high-speed data transmission today. The textbook on data communications that he co-authored became the most cited reference in the communications field over the period of a decade. At Bell Labs, Dr. Lucky became executive director of the Communications Sciences Research Division in 1982, where he was responsible for research on the methods and technologies for future communication systems. In 1992, he left Bell Labs to assume his present position at Bellcore. He has served as president of the Communications Society of the IEEE, and as vice president and executive vice president of the parent IEEE. He has been editor of several technical journals, including the *Proceedings of the IEEE,* and, since 1982, he has written the bi-monthly "Reflections" column of personalized observations about the engineering profession in *Spectrum* magazine. In 1993, these "Reflections" columns were collected in the IEEE Press book, *Lucky Strikes...Again.*

Dr. Lucky is a fellow of the IEEE and a member of the National Academy of Engineering. He is also a consulting editor for a series of books on communications through Plenum Press. He has been on the advisory boards or committees of many universities and government organizations and was chairman of the Scientific Advisory Board of the U.S. Air Force from 1986-1989. He was the 1987 recipient of the prestigious Marconi Prize for his contributions to data communications and has been awarded honorary doctorates from Purdue University and the New Jersey Institute of Technology. He has also been awarded the Edison Medal of the IEEE and the Exceptional Civilian Contributions Medal of the U.S. Air Force. Dr. Lucky has been an invited lecturer at about 100 different universities, and has been the guest on a number of network television shows, including Bill Moyers' "A World of Ideas," where he has discussed the impacts of future technological advances. He is the author of the popular book, *Silicon Dreams,* which is a semi-technical and philosophical discussion of the ways in which both

APPENDIX C

humans and computers deal with information. Dr. Lucky served on or advised the committees that produced the following CSTB reports: *Keeping the U.S. Computer Industry Competitive: Defining the Agenda; Keeping the U.S. Computer Industry Competitive: Systems Integration; Keeping the U.S. Computer and Communications Industry Competitive: Convergence of Computing, Communications, and Entertainment; Realizing the Information Future: The Internet and Beyond*; and *The Unpredictable Certainty: Information Infrastructure Through 2000*.

JOHN MAJOR, QUALCOMM, Inc.

John Major, a current CSTB member, at the time of the symposium was the senior vice president and assistant chief corporate staff officer for Motorola. One of his responsibilities was leading Motorola's initiative to become a global leader in software. Previously, he managed the Worldwide Systems Group that developed and manufactured private radio systems for voice and data for public safety and business users. Mr. Major holds a B.S. in mechanical and aerospace engineering from the University of Rochester, an M.S. in mechanical engineering from the University of Illinois, an M.B.A. with distinction from Northwestern University, and a J.D. from Loyola University. He serves as chairman of the board of directors of the Telecommunications Industry Association (TIA), and he serves on the board of directors of the Electronics Industry Association (EIA). Mr. Major also serves on the board of directors of Littelfuse and Lennox Corporation, and is trustee of the Allendale School, which helps disadvantaged children. He currently chairs the board of health for Barrington Hills.

ROBERT M. METCALFE, International Data Group

Robert Metcalfe, a founding member of CSTB, is executive correspondent, *INFOWORLD*, and vice president/ technology, International Data Group. He was born in 1946 in Brooklyn, New York, and grew up on Long Island. He graduated in 1969 after five years at MIT, receiving a bachelor's degree in electrical engineering and a bachelor's degree from the Sloan School of Management. In 1970, Bob received a master's degree in applied mathematics from Harvard University. In 1973, he received a Ph.D. from Harvard in computer science for research done at MIT's project Mac on packet switching in the DARPA and Aloha computer networks. In 1972, Dr. Metcalfe moved to the Computer Science Laboratory at the Xerox Palo Alto Research Center (PARC) to join in the early development of personal computing. In 1973, he invented Ethernet, the local-area networking technology on which he shares four patents.

While at PARC, he began eight years of part-time teaching at Stanford University, finishing in 1983 as a consulting associate professor of electrical engineering with a new course on distributed computing. In 1976, Metcalfe moved to Xerox's Systems Development Division to manage microprocessor and communication developments that led, long after he left, to the Xerox Star workstation. Metcalfe left Xerox in 1979 to promote personal computer local-area networks (PC LANs) and, especially, Ethernet. Also in 1979, Metcalfe founded 3Com Corporation, the Fortune 500 computer networking company where he held various positions, including chairman of the board of directors, chief executive officer, president, vice president of engineering, vice president of sales and marketing, chief technical officer, and general manager consecutively of the software, workstation, and hardware divisions. Metcalfe retired from 3Com in 1990 after 11 years.

In June 1996, Dr. Metcalfe was awarded the IEEE Medal of Honor for his exemplary and sustained leadership in the development, standardization, and commercialization of Ethernet. In 1980, he received the Grace Murray Hopper Award from the ACM, and, in 1988, the Alexander Graham Bell Medal from the IEEE—both for his invention, standardization, and commercialization of local-area networks. Metcalfe's many publications include *Packet Communication*, his groundbreaking Harvard Ph.D. dissertation published in book form in 1996; the often-cited *Ethernet: Distributed Packet Switching for Local Computer Networks*, with David Boggs in the *Communications of the ACM*, July 1976; and *Local Networks of Personal Computers, at the Ninth World Computer Congress* in Paris in 1983.

Dr. Metcalfe served for a year on the Executive Office of the President's Advisory Committee on Information Networks. For two years he was chairman of the Corporation for Open Systems, promoting worldwide computer and telephone networking standards. In 1991-92, Dr. Metcalfe was a visiting fellow at Wolfson College in the computer laboratory of the University of Cambridge, England. Dr. Metcalfe was conference chair for ACM97: The Next 50 Years of Computing, San Jose Convention Center, March 1-5, 1997.

DAVID B. NELSON, Department of Energy

David Nelson is associate director of the Office of Energy Research for Computational and Technology Research in the U. S. Department of Energy. He manages programs that include technology research, technology transfer, mathematics, computer and computational science, the Energy Sciences data network (ESnet), and several supercomputer centers. He also serves as the Office of Energy Research member of the departmental standards committee, which is overseeing significant changes in the Department's approach to environment, safety, and health. Previously, he served as associate director of energy research, providing general assistance to the director of energy research. Specific responsibilities included: oversight of environment, safety, and health; technology transfer and industrial cooperation; computing, telecommunications, and information; and termination of the Superconducting Super Collider. He is past chairman of the National Science and Technology Council Subcommittee on High Performance Computing, Communications, and Information Technology, a multi-agency planning and coordinating body of the federal government organized under the President's Science Advisor. He is the Department's alternate representative on the administration's Information Infrastructure Task Force, organized under the National Economic Council. Dr. Nelson moved to the Department of Energy in 1979 from the Oak Ridge National Laboratory (ORNL), where he was a research scientist working mainly in theoretical plasma physics and its applications to fusion energy, and also in defense and environmental research. He headed the magneto-hydrodynamics theory group in the fusion energy division.

Dr. Nelson received his A.B. cum laude from Harvard University, majoring in engineering sciences. After working as an electrical engineer in New York City, he studied at the Courant Institute of Mathematical Sciences at New York University, where he received the M.S. and Ph.D. degrees, both in mathematics. He received additional graduate training in mathematics at the Freie Universitat, Berlin, and in physics at Columbia University and the University of Tennessee. In 1975-76, he returned to the Courant Institute as visiting member on leave from ORNL. He is the author of numerous papers in theoretical plasma physics, computational science, and research policy, and is a member of the American Physical Society.

MICHAEL NELSON, Federal Communications Commission

Michael Nelson at the time of the symposium was special assistant for information technology at the White House Office of Science and Technology Policy. He has worked closely with both Jack Gibbons, the President's Science Advisor, and Vice President Gore on a wide range of issues relating to the National Information Infrastructure (NII), including telecommunications policy, high-performance computing, encryption, and information policy. He has been part of the Information Infrastructure Task Force, which is responsible for coordinating the administration's NII Initiative, and has worked closely with the Vice President on the administration's new Global Information Infrastructure initiative to link together national and international networks in a seamless "network of networks." He is now responsible for technology policy at the Federal Communications Commission.

Prior to moving to the White House, Dr. Nelson served for five years on the staff of the Senate Commerce Committee, where he worked closely with then-Senator Gore, the chairman of the Subcommittee on Science, Technology, and Space. Among the issues he handled were global warming, high-performance computing research, earthquake issues, Antarctica, and biotechnology. He was the lead Senate staffer on Gore's High-Performance Computing Act of 1991, which authorized the $1-billion HPCC Program that is helping to develop the technologies needed for the NII. Dr. Nelson received a B.S. in geology from the California Institute of Technology in 1981 and a Ph.D. in geophysics from MIT in 1988.

RAJ REDDY, Carnegie Mellon University

Raj Reddy, a founding member of CSTB, is dean of the School of Computer Science at Carnegie Mellon University and the Herbert A. Simon University professor of computer science and robotics. Dr. Reddy joined

APPENDIX C

Carnegie Mellon's department of computer science in 1969 and served as director of the Robotics Institute from 1979 to 1992. Previously, he was an assistant professor of computer science at Stanford University from 1966 to 1969, and served as an applied science representative for IBM in Australia from 1960 to 1963. His research interests include the study of human-computer interaction and artificial intelligence. His current research projects include speech recognition and understanding systems; multi-media digital libraries; just-in-time learning technologies; and the automated machine shop project. His professional honors include: fellow of IEEE, ASA, and AAAI; member of the National Academy of Engineering; president of AAAI, 1987-89; IBM Research Ralph Gomory Fellow, 1991; and the Turing Award, 1994. Dr. Reddy was presented the Legion of Honor by President Mitterrand of France in 1984. He served on the committees that produced the CSTB reports, *Computing the Future: A Broader Agenda for Computer Science and Engineering* and *Information Technology for Manufacturing*.

CHARLES L. SEITZ, Myricom, Inc.

Charles Seitz, vice chair of CSTB at the time of the symposium, is the president of Myricom, Inc., a startup company involved in research, development, production, and sales of high-speed computers and local-area networks. During the 16 years prior to founding Myricom, he was a professor of computer science at the California Institute of Technology, where his research and teaching were in the areas of VLSI design, computer architecture and programming, and concurrent computation. He earned S.B. (1965), S.M. (1967) and Ph.D. (1971) degrees from MIT, where he was also an instructor and the recipient of the Goodwin Medal "for conspicuously effective teaching." He was a consultant and member of the technical staff of the Evans & Sutherland Computer Corporation during its initial years (1968-72), an assistant professor of computer science at the University of Utah (1970-72), and a consultant and leader of several research and development projects for Burroughs Corporation (1971-78). His research in VLSI and concurrent computing at Caltech, including the development of the Cosmic Cube multicomputer, was selected by *Science Digest* as one of the top 100 innovations in 1985. Dr. Seitz was elected to the National Academy of Engineering in 1992 "for pioneering contributions to the design of asynchronous and concurrent computer systems." Dr. Seitz served on the committee that produced the CSTB report, *Computing the Future: A Broader Agenda for Computer Science and Engineering*.

MARY SHAW, Carnegie Mellon University

Mary Shaw, a founding member of CSTB, is the Alan J. Perlis professor of computer science, associate dean for professional programs, and member of the Human Computer Interaction Institute at Carnegie Mellon University. She has been a member of this faculty since completing the Ph.D. degree at Carnegie Mellon in 1972. From 1984 to 1987, she served as chief scientist of CMU's Software Engineering Institute. She had previously received a B.A (cum laude) from Rice University and worked in systems programming and research at the Research Analysis Corporation and Rice University.

Her research interests in computer science lie primarily in the areas of programming systems and software engineering, particularly software architecture, programming languages, specifications, and abstraction techniques. Particular areas of interest and projects have included software architectures (Vitruvius, UniCon), technology transition (SEI), program organization for quality human interfaces (Descartes), programming language design (Alphard, Tartan), abstraction techniques for advanced programming methodologies (abstract data types, generic definitions), reliable software development (strong typing and modularity), evaluation techniques for software (performance specification, compiler contraction, software metrics), and analysis of algorithms (polynomial derivative evaluation). She has developed innovative curricula from the introductory to the doctoral level.

Dr. Shaw is an author or editor of seven technical books and more than 100 papers and reports. In 1993, she received the Warnier prize for contributions to software engineering. She is a fellow of the ACM, IEEE, and AAAS. She is also a member of the Society of the Sigma Xi, the New York Academy of Sciences, and Working Group 2.4 (System Implementation Languages) of the International Federation of Information Processing Societies. In addition, she has served on a number of advisory and review panels, conference program committees, and editorial boards. Dr. Shaw served on the committee that produced the CSTB report, *Scaling Up: A Research Agenda for Software Engineering*.

EDWARD H. SHORTLIFFE, Stanford University

Edward Shortliffe, a CSTB member at the time of the symposium, is professor of medicine and of computer science at Stanford University. He received an A.B. in applied mathematics from Harvard College in 1970, a Stanford Ph.D. in medical information sciences in 1975, and an M.D. at Stanford in 1976. During the early 1970s, he was principal developer of the medical expert system known as MYCIN. After a pause for internal medicine house-staff training at Harvard and Stanford between 1976 and 1979, he joined the Stanford internal medicine faculty, where he has directed an active research program in clinical information systems development. His interests include the broad range of issues related to integrated decision-support systems and their effective implementation. He has spearheaded the formation of a Stanford degree program in medical informatics, and continues to divide his time between clinical medicine and medical-informatics research. He is currently associate dean for information technology at Stanford University School of Medicine.

Dr. Shortliffe is a member of the Institute of Medicine, the American Society for Clinical Investigation, the Association of American Physicians, and the American Clinical and Climatological Association. He has also been elected to fellowship in the American College of Medical Informatics, the American Association for Artificial Intelligence, and the American College of Physicians. He sits on the editorial boards of several medical computing and artificial intelligence publications. He has served on the Federal Networking Advisory Committee, the Biomedical Library Review Committee, and was recipient of a research career development award from the latter agency. In addition, he received the Grace Murray Hopper Award of the Association for Computing Machinery in 1976 and has been a Henry J. Kaiser Family Foundation Faculty Scholar in General Internal Medicine. Dr. Shortliffe has authored over 150 articles and books in the fields of medical computing and artificial intelligence. Volumes include *Computer-Based Medical Consultations: MYCIN* (Elsevier/North Holland, 1976), *Readings in Medical Artificial Intelligence: The First Decade* (with W.J. Clancey; Addison-Wesley, 1984), *Rule-Based Expert Systems: The MYCIN Experiments of the Stanford Heuristic Programming Project* (with B.G. Buchanan; Addison-Wesley, 1984), and *Medical Informatics: Computer Applications in Health Care* (with L.E. Perreault, G. Wiederhold, and L.M. Fagan; Addison-Wesley, 1990). Dr. Shortliffe co-chaired the CSTB planning session on "The Roles of Information Infrastructure in Health and Health Care."

WILLIAM J. SPENCER, SEMATECH

Since October 1990, William Spencer, a former member of CSTB, has been president and chief executive officer of SEMATECH in Austin, Texas. SEMATECH is a research and development consortium jointly funded by semiconductor industry member companies and the U.S. government, established to solve the technical challenges required to keep the U.S. number one in the global semiconductor industry. Before joining SEMATECH, Dr. Spencer was group vice president and senior technical officer at Xerox Corporation in Stamford, Connecticut. He has also served as vice president of Xerox Palo Alto Research Center, director of systems development at Sandia National Laboratories in Livermore, and director of microelectronics at Sandia National Laboratories in Albuquerque. He began his career at Bell Telephone Laboratories. Dr. Spencer received an A.B. degree from William Jewell College in Liberty, Missouri, followed by an M.S. degree in mathematics and a Ph.D. in physics from Kansas State University. He was awarded the Regents Meritorious Service Medal from the University of New Mexico in 1981, and an honorary doctorate degree from William Jewell College in 1990. He is a member of the National Academy of Engineering, a fellow of IEEE, and serves on numerous advisory groups and boards.

JOSEPH F. TRAUB, Columbia University

Joseph Traub, founding chair of CSTB, is the Edwin Howard Armstrong professor of computer science at Columbia University and external professor at the Santa Fe Institute. From 1971 to 1979, he was head of the computer science department at Carnegie Mellon University. Beginning in 1959, Dr. Traub pioneered research in what is now called information-based complexity, which studies the computational complexity of problems with partial or contaminated information. His current work ranges from new fast methods for pricing financial derivatives to investigating what is scientifically knowable. As part of the latter, he directs a center for the study of limits to scientific knowledge at the Santa Fe Institute, partially funded by the Alfred P. Sloan Foundation.

Dr. Traub is the author or editor of eight books and some 100 journal articles. He is the founding editor of the *Journal of Complexity*. A Festschrift in celebration of his 60th birthday was recently published. Dr. Traub has received numerous honors, including election to the National Academy of Engineering (1985), and he is a fellow of both the AAAS (1972) and the ACM (1993). He received the 1991 Emanuel R. Piore Gold Medal Award of IEEE and the 1992 Distinguished Service Award of the Computing Research Association. He has been Sherman Fairchild distinguished scholar at the California Institute of Technology and received a Distinguished Senior Scientist Award from the Alexander von Humboldt Foundation. He was selected by the Accademia Nazionale dei Lincei in Rome to present the 1993 Lezioni Lincei, a cycle of six lectures that will be published by Cambridge University Press. He has served as advisor or consultant to the senior management of numerous organizations including IBM, Hewlett-Packard, Schlumberger, Stanford University, INRIA (Paris), the Federal Judiciary Center, DARPA, and NSF. Dr. Traub served on the committee that produced the CSTB report, *Evolving the High Performance Computing and Communications Initiative to Support the Nation's Information Infrastructure*.

DR. ANDRIES van DAM, Brown University

Andries van Dam, a founding member of CSTB, has been on the faculty of Brown University since 1965. He was one of the founders of the department of computer science, and its first chairman from 1979 to 1985. He is also director of the NSF/DARPA National Science and Technology Center for Graphics and Visualization, a research consortium including Brown University, California Institute of Technology, Cornell University, University of North Carolina, and Utah University. His research has concerned computer graphics, text processing and hypermedia systems, and workstations. He has been working for nearly 30 years on systems for creating and reading "electronic books" with interactive illustrations, based on high-resolution interactive graphics systems, for use in teaching and research. Most recently, he has been concerned with teaching object-oriented programming and design to entering students. Dr. van Dam received the B.S. degree with Honors from Swarthmore College in 1960 and the M.S. and Ph.D. degrees from the University of Pennsylvania in 1963 and 1966, respectively. A member of Sigma Xi, IEEE Computer Society, and ACM, he helped to found, and from 1971 to 1981 was an editor of, *Computer Graphics and Image Processing*, and was an editor of ACM's *Transactions on Graphics* from 1981 to 1986. He became a member of the editorial board of the IEEE's *Transactions on Visualization and Computer Graphics*. In 1967, Professor van Dam co-founded ACM's SIGGRAPH.

In 1974, Dr. van Dam received the Society for Information Display's "Special Recognition Award," and in 1984 the IEEE Centennial Medal. In 1988, he received the state of Rhode Island Governor's Science and Technology Award, and in 1990 he received the National Computer Graphics Association's Academic Award. In July 1991, he received SIGGRAPH's Steven A. Coons Award. In May 1992, Brown University named him to the L. Herbert Ballou university professor chair and in March of 1995 to the Thomas J. Watson, Jr., university professor of technology and education chair. In 1994, he received the Karl V. Karlstrom Outstanding Educator Award, the IEEE Fellow Award, and the ACM Fellow Award. In December 1995, he received an honorary Ph.D. from Darmstadt Technical University in Germany. In June 1996, he received an honorary Ph.D. from his alma matter, Swarthmore College. He is past chairman of the Computing Research Association, a founder and chief scientist of Electronic Book Technologies, a member of the technical advisory boards for Object Power, Inc., the Fraunhofer Center for Computer Graphics, and Microsoft. In 1996, he was elected to the National Academy of Engineering for his contributions to computer science education and graphics research.

PAUL R. YOUNG, University of Washington

Paul Young at the time of the symposium was the National Science Foundation's assistant director for the Directorate of Computer and Information Science and Engineering (CISE). Dr. Young was responsible for 25 programs organized into six divisions representing the areas of Computer and Computation Research; Information, Robotics and Intelligent Systems; Advanced Scientific Computing; Microelectronic Information Processing Systems; Networking and Communications Research and Infrastructure; and Cross-Disciplinary Activities. Before joining the National Science Foundation, Dr. Young was professor of computer science and engineering and associate dean of engineering at the University of Washington. He is a graduate of Antioch College and received his Ph.D. from MIT in 1963. He joined the University of Washington in 1983, after 17 years at Purdue University, where he was one of the first half dozen faculty members in perhaps the first computer science department in the United States.

Dr. Young has been a Brittingham visiting professor in computer science at the University of Wisconsin, has twice taught as a visiting professor in the electrical engineering and computer sciences department at the University of California, Berkeley, served briefly as chairman of the Computing and Information Sciences Department at the University of New Mexico, and has been a National Science Foundation postdoctoral fellow at Stanford University. His research interests are in theoretical computer science, with an emphasis on questions of computational complexity and on connections with mathematical logic. He is author or co-author of some 36 research papers and more than a half dozen expositories in this area, and is co-author of a graduate textbook on the general theory of algorithms.

Dr. Young served on the board of directors of the Computing Research Association from 1983 to 1991 and was chairman from 1989-91. From 1983-1988, he served as chairman of the Computer Science Department at the University of Washington. In 1977-80, he served on the National Science Foundation's Advisory Subcommittee for Computer Science and served as chairman of this subcommittee in 1979-80. Dr. Young has served three times on the program committee for the ACM's Symposium on the Theory of Computing, and he has served on both the executive committee and the nominating committee for ACM's Special Interest Group on the Theory of Computing. He has also been chairman of the program committee for the IEEE Computer Society's Annual Symposium on the Foundations of Computer Science, and he has served as both vice chairman and chairman of the Computer Society's Technical Committee on the Mathematical Foundations of Computing. He has also served on the program committee, and later as chair of the program committee, for the IEEE Structural Complexity Theory Conference. In 1995, he was elected a fellow of both the IEEE and the ACM. Dr. Young served on the committee that produced the CSTB report, *Computing Professionals: Changing Needs for the 1990s*.